FOOD SCIENCE

Research and Technology

FOOD SCIENCE

Research and Technology

Edited By
A. K. Haghi, PhD
Associate member of University of Ottawa, Canada;
Freelance Science Editor, Montréal, Canada

Apple Academic Press
TORONTO NEW JERSEY

Apple Academic Press Inc.
3333 Mistwell Crescent
Oakville, ON L6L 0A2
Canada

Apple Academic Press Inc.
1613 Beaver Dam Road, Suite # 104
Point Pleasant, NJ 08742
USA

Exclusive worldwide distribution by CRC Press, a Taylor & Francis Group

International Standard Book Number: 978-1-926895-01-7 (Hardback)

Library and Archives Canada Cataloguing in Publication

Food science: research and technology/edited by A.K. Haghi.

Includes bibliographical references and index.
ISBN 978-1-926895-01-7
1. Food–Analysis. 2. Food–Composition. 3. Food industry and trade. I. Haghi, A. K.

TX541.F66 2011 664'.07 C2011-906633-5

Apple Academic Press also publishes its books in a variety of electronic formats. Some content that appears in print may not be available in electronic format. For information about Apple Academic Press products, visit our website at **www.appleacademicpress.com**

Contents

List of Contributors

Anna Ardévol
Departament de Bioquímica i Biotecnologia, Universitat Rovira i Virgili (URV), Tarragona, Spain.

Muhammad Ashraf
Department of Botany, Faculty of Sciences, University of Agriculture, Faisalabad 38040, Pakistan.

Mário C. J. Bier
Bioprocess Engineering and Biotechnology Department, Federal University of Paraná (UFPR), CEP 81531-990, Curitiba–PR, Brazil.

Mayte Blay
Departament de Bioquímica i Biotecnologia, Universitat Rovira i Virgili (URV), Tarragona, Spain.

Elizabeth Carvajal–Millán
Laboratorio de biopolímeros, CTAOA, Centro de Investigación en Alimentación y Desarrollo, A.C. (CIAD). Carretera a La Victoria, P.O. Box 1735, Hermosillo, Sonora 83000, Mexico.

Karla Escárcega-Loya
Cuauhtémoc, Chihuahua, México, Centro de Investigación en Alimentación y Desarrollo, A.C. (CIAD). Carretera a La Victoria, P.O. Box 1735, Hermosillo, Sonora 83000, Mexico.

Leifa Fan
Institute of Horticulture, Zhejiang Academy of Agricultural Sciences, 139 Shiqiao Road, Hangzhou-ZJ, P. R. China.

Guillermina García-Sánchez
CTAOA, Centro de Investigación en Alimentación y Desarrollo, A.C. (CIAD). Carretera a La Victoria, P.O. Box 1735, Hermosillo, Sonora 83000, Mexico.

Alfonso Gardea
CIAD Guaymas, Col. Las Playitas 85480, Guaymas, Sonora, México, Centro de Investigación en Alimentación y Desarrollo (CIAD), A.C. Carretera a La Victoria, P.O. Box 1735, Hermosillo, Sonora 83000, Mexico.

Sascha Habu
Department of Bioprocess and Biotechnology, Federal University of Paraná, Rua Francisco H. dos SantosCentro Politécnico, Jardim das AméricasCuritiba/PR—Brazil.

Akbar K. Haghi
University of Guilan, Rasht, Iran.

Alma Rosa Islas
Laboratorio de biotecnología. CTAOV, Centro de Investigación en Alimentación y Desarrollo, A.C. (CIAD). Carretera a La Victoria, P.O. Box 1735, Hermosillo, Sonora 83000, Mexico.

Amin Ismail
Department of Nutrition and Dietetics, Faculty of Medicine and Health Sciences, Universiti Putra Malaysia, 43400 UPM Serdang, Selangor, Malaysia.

Yueming Jiang
Key Laboratory of Plant Resources Conservation and Sustainable Utilization of Chinese Academy of Sciences, South China Botanical Garden, Chinese Academy of Sciences, Guangzhou 510650, China.

Jaime Lizardi
Laboratorio de biopolímeros, CTAOA, Centro de Investigación en Alimentación y Desarrollo, A.C. (CIAD). Carretera a La Victoria, P.O. Box 1735, Hermosillo, Sonora 83000, Mexico.

Jorge Márquez-Escalantle
Cuauhtémoc, Chihuahua, Centro de Investigación en Alimentación y Desarrollo, A.C. (CIAD). Carretera a La Victoria, P.O. Box 1735, Hermosillo, Sonora 83000, Mexico.

Adriane B. P. Medeiros
Bioprocess Engineering and Biotechnology Department, Federal University of Paraná (UFPR), CEP 81531-990, Curitiba–PR, Brazil.

Nornaemah M. B.
Universiti Sains Malaysia, 11800 Gelugor, Pulau Pinang, Malaysia.

Nor Hafizah N.
Universiti Sains Malaysia, 11800 Gelugor, Pulau Pinang, Malaysia.

Hasnah O.
Universiti Sains Malaysia, 11800 Gelugor, Pulau Pinang, Malaysia.

Montserrat Pinent
Departament de Bioquímica i Biotecnologia, Universitat Rovira i Virgili (URV), Tarragona, Spain.

K. Nagendra Prasad
Key Laboratory of Plant Resources Conservation and Sustainable Utilization of Chinese Academy of Sciences, South China Botanical Garden, Chinese Academy of Sciences, Guangzhou 510650, China.
Department of Nutrition and Dietetics, Faculty of Medicine and Health Sciences, Universiti Putra Malaysia, 43400 UPM Serdang, Selangor, Malaysia.

Agustín Rascón–Chu
Laboratorio de biotecnología. CTAOV, Centro de Investigación en Alimentación y Desarrollo, A.C. (CIAD). Carretera a La Victoria, P.O. Box 1735, Hermosillo, Sonora 83000, Mexico.

Afidah A. R.
Universiti Sains Malaysia, 11800 Gelugor, Pulau Pinang, Malaysia.

Rizal R.
Herbal Village Industries, 06250, Alor Setar, Kedah, Malaysia.

Alejandro Romo-Chacón
Cuauhtémoc, Chihuahua, Centro de Investigación en Alimentación y Desarrollo, A.C. (CIAD). Carretera a La Victoria, P.O. Box 1735, Hermosillo, Sonora 83000, Mexico.

Suzan C. Rossi
Bioprocess Engineering and Biotechnology Department, Federal University of Paraná (UFPR), CEP 81531-990, Curitiba–PR, Brazil.

Carlos R. Soccol
Bioprocess Engineering and Biotechnology Department, Federal University of Paraná (UFPR), CEP 81531-990, Curitiba–PR, Brazil.

Nurulafiqah M. T.
Universiti Sains Malaysia, 11800 Gelugor, Pulau Pinang, Malaysia.

Vania Urias–Orona
Laboratorio de biopolímeros, CTAOA, Centro de Investigación en Alimentación y Desarrollo, A.C. (CIAD). Carretera a La Victoria, P.O. Box 1735, Hermosillo, Sonora 83000, Mexico.

Luciana P. S. Vandenberghe
Bioprocess Engineering and Biotechnology Department, Federal University of Paraná (UFPR), CEP 81531-990, Curitiba–PR, Brazil.

List of Abbreviations

AAT	Alcohol acetyl-transferase
Acetyl-CoA	Acetyl coenzyme A
BHA	Butylated hydroxyanisole
BHT	Butylated hydroxytoluene
BI	Browning index
bw	Body weight
DE	Degree of esterification
DM	Degree of methoxylation
DPPH	2,2-Diphenyl-1-picrylhydrazyl
FDA	Food and Drug Administration
FF	Functional food
FRAP	Ferric reducing antioxidant power
FTIR	Fourier transform infrared
G6pc	Glucose-6-phosphatase
Gck	Glucokinase
Gckr	Glucokinase-regulatory protein
Glut2	Glucose transporter type 2
Glut4	Glucose transporter type 4
GO	Galactooligosaccharides
GSPE	Grape seed procyanidin extract
HCA	Hydroxycitric acid
Hk2	Hexokinase 2
HM	High methoxy
HPE	High pressure extraction
HPLC	High performance liquid chromatography
IC	Inhibitory concentration
LFP	Litchi fruit pericarp
LM	Low methoxy
MPD	Mean polymerization degree
MRSA	Methicillin-resistant *Staphylococcus aureus*
M*v*	Viscosimetric molecular weight
NDOs	Nondigestible oligosaccharides
ORAC	Oxygen radical absorbance capacity
PCMB	p-Chloromercuribenzoato
qNPGal	q-Nitrophenyl-α-D-galactopyranoside

ROs	Raffinose oligosaccharides
ROS	Reactive oxygen species
STZ	Streptozotocin
TFC	Total flavonoid content
TMCA	3,4,5-Trimethoxy-trans-cinnamic acid
TPC	Total phenolic content
TPTZ	2,4,6-Tripyridyl-s-triazine
TTC	Total tannin content
UE	Ultrasonic extraction
UV-Vis	Ultraviolet-visible
WHO	World Health Organization

Preface

The collection of topics in this book aims to reflect the diversity of recent advances in Food Science and Technology with a broad perspective which may be useful for scientists as well as for graduate students and engineers. This new book presents leading-edge research from around the world in this dynamic field.

Diverse topics published in this book are the original works of some of the brightest and well-known international scientists.

The book offers scope for academics, researchers, and engineering professionals to present their research and development works that have potential for applications in several disciplines of Food Science and Technology. Contributions ranged from new methods to novel applications of existing methods to gain understanding of the material and/or structural behavior of new and advanced systems.

Contributions are sought from many areas of food science and technology in which advanced methods are used to formulate (model) and/or analyze the problem. In view of the different background of the expected audience, readers are requested to focus on the main ideas, and to highlight as much as possible the specific advantages that arise from applying modern ideas. A chapter may therefore be motivated by the specific problem, but just as well by the advanced method used which may be more generally applicable.

I would like to express my deep appreciation to all the authors for their outstanding contribution to this book and to express my sincere gratitude for their generosity. All the authors eagerly shared their experiences and expertise in this new book. Special thanks go to the referees for their valuable work.

— A. K. Haghi

Chapter 1

Microbial Enzymes for Flavor, Dairy, and Baking Industries

Adriane B. P. Medeiros, Suzan C. Rossi, Mário C. J. Bier, Luciana P. S. Vandenberghe, and Carlos R. Soccol

INTRODUCTION

Microbial enzymes have been exploited commercially over the years. They are of great significance and have greatly expanded demand in the food industry. It has been reported that microbial enzymes account for 45% of the global food enzymes sales. Because of improved understanding of production biochemistry, the fermentation processes, and recovery methods, an increasing number of enzymes can be optimized for existing applications. This chapter focuses mainly on the types of enzymes, and the possible applications of these enzymes in the dairy, baking, and flavoring for food.

Enzymes from different sources (plants, animals, and microorganisms) have wide application in different sectors. Microbial enzymes are often more stable and useful than enzymes derived from plants or animals because of the great variety of catalytic activities available, the high yields possible, ease of genetic manipulation, regular supply due to absence of seasonal fluctuations, and rapid growth of microorganisms on inexpensive media (Hasan et al., 2005). According to a report from Business Communications Company, Inc. the global market for industrial enzymes increased from $2.2 billion in 2006 to an estimated $2.3 billion by the end of 2007. It should reach $2.7 billion by 2012, a compound annual growth rate of 4%. The total industrial enzyme market in 2009 is expected to reach nearly $2.4 billion (Thakore, 2008). Food industry is the largest consumer of enzymes, and approximately 45% of bulk share goes to it (Pandey and Ramachandran, 2005).

Enzymes are commonly used in food processing and in the production of food ingredients. As indicated in Table 1.1, applications of enzymes in the food industry are many and diverse. Including the production of food quality attributes such as flavors, control of color, texture, and appearance. Several advances have been made in optimization enzymes for existing applications and in the use of recombinant protein production to provide efficient mono-component enzymes that do not result in harmful side effects (Kirk et al., 2002).

Table 1.1. Applications of enzymes in the food industry.

Enzyme	Application
Protease	Milk clotting, infant formulas (low allergenic), biscuits, cookies, flavor
Lipase	Cheese flavor, dough stability
Lactase	Lactose removal (milk)
Pectin methyl esterase	Firming fruit-based products

Source: Adaped from Kirk et al. (2002).

ENZYMES INVOLVED IN AROMA PRODUCTION

The aroma formulator consists in constructing a flavor recalling a true and original aroma in a processed food product with a specific texture and composition. To be able to formulate this flavor, a technologist needs a wide spectrum of different aroma components. These compounds can be extracted from fruits or vegetables but, as they are required in the product in concentrations comparable to those in the source material, this utilizes high amounts of materials and is generally not economically realistic. Most food flavoring compounds can also be produced via chemical synthesis. However, recent market surveys have demonstrated that consumers prefer foodstuff that can be labeled "natural". The use of biotechnology appears to be attractive in various ester preparations under milder conditions, and the product may be given the natural label (Salah et al., 2007).

Biotechnology proposes to use enzymes or whole cells to produce aroma compounds from natural substrates in a way often inspired by biochemical pathways encountered in nature. The challenge is to put a naturally rich source of substrate in contact with highly active enzymes. In adequate conditions, this can result in the production of flavor compounds in mass fractions of the order of several g/kg, instead of mg/kg encountered in raw materials. The resulting flavor compounds are called natural since they are produced from agro-products through natural biological activities (Aguedo et al., 2004). Besides the advantage of being recognized as natural, the aromas produced by microorganisms have a great economic potential for obtaining a wide range of bio-molecules of interest in the food industry (Rossi et al., 2009; Soccol et al., 2008). The industrial production of aromas corresponds of 25% of food additives world market (Couto and Sanromán, 2006). The world market of flavors and fragrances has a current volume of US $18.6 billion (Soccol et al., 2007). Still 10% of the supply is derived from bioprocesses. Examples, such as the Bartlett pear impact compound, ethyl 2, 4-(E, Z)-deca-dienoate, which is cheaper to produce using enzyme catalysis than chemosynthesis, should encourage further research (Berger, 2009).

The majority method for the production of natural aromas has enzymes involved in biosynthesis. A number of enzymes (lipases, proteases, and glycosidases) catalyzes the production of aroma-related compounds from precursor molecules. Despite the higher costs involved, microbial enzymes—rather than cells—can offer high stereo- and enatio-selectivity toward substrate conversion techniques such as enzyme immobilization, and eventually coenzyme regeneration might result in highly efficient and specific bio-catalytic processes for flavor synthesis (Soccol et al., 2008; Vandamme and Soetaert, 2002).

Lipases

Among the hydrolytic enzymes of interest are the lipases (triacyl-glycerohydrolases, EC 3.1.1.3). These enzymes had been used for the hydrolysis of acyl-glycerides, which are versatile biocatalysts that can catalyze different reactions in both aqueous and organic media, with limited water content. Among the lipases of plant, animal, and microbial, these are the most used, and, mostly, are not harmful to human health, and are recognized as "Generally Regarded as Save—GRAS" (Olempska-Beer et al., 2006).

From the industrial point of view, the fungi is especially valued because the enzymes they produce are usually extracellular, which facilitates their recovery from the fermentation medium.

Lipases have been used in a variety of biotech sectors, as in the food (flavor development and cheese ripening), detergent, oleo chemical (hydrolysis of fats and oils, synthesis of bio-surfactants), and for treatment of oily wastes emanating from the leather and paper (Carvalho et al., 2005). This enzyme mediated synthesis of flavor esters has the potential of satisfying the increasing demand for natural flavor esters. Enzymatic esterification of flavor esters has the advantage of catalyzing reactions more specifically than chemical synthesis under mild condition and higher yield over microbiological production.

In recent years, lipases have been found to catalyze reversible reactions such as esterifying reaction in aqueous, non-aqueous, and solvent-free phases. Lipase-mediated synthesis of aliphatic esters of longer chain substrates has shown the potential of their esterification abilities. The synthesis of low molecular flavor ethyl esters from shorter chain substrate has been attempted. However, this attempt comparatively received less attention with no satisfaction as short fatty-acids easily strip the essential water around enzymes to cause their de-activation or cause dead-end inhibition reacting with the serine residue at the active site of lipase. Almost all researchers did not succeed in synthesizing them with high yields (>80%) at more than 0.5 M substrate concentration, which keeps enzymatic pathway for synthesis of ethyl esters of short-chain fatty acids far from industries (Xu et al., 2002).

The lipases, both free and immobilized, have a high denature capability. Strains capable of producing active and stable lipases to catalyze esterification of ethanol with short-chain fatty acids in a non-aqueous phase, should be screened for the synthesis of flavor esters. Lipases have been used for addition to food to modify flavor by synthesis of esters of short chain fatty acids and alcohols, which are known flavor and fragrance compounds because psychotropic gram-negative bacteria, such as *Pseudomonas* species, pose a significant spoilage problem in refrigerated meat and dairy products due to secretion of hydrolytic enzymes, especially lipases and proteases.

Lipases have earlier been used in production of leaner meat such as in fish. The fat is removed during the processing of the fish meat by adding lipases and this procedure is called bio-lipolysis. The lipases also play an important role in the fermentative steps of sausage manufacture and to determine changes in long-chain fatty acid liberated during ripening. Earlier, lipases of different microbial origin have been used for refining rice flavor, modifying soyabean milk, and for improving the aroma and accelerating the fermentation of apple wine (Hasan et al., 2005). These are of great importance for the dairy industry for the hydrolysis of milk fat. Present applications include the flavor enhancement of cheeses, the acceleration of cheese ripening, the manufacturing of cheese like products, and the lipolysis of butterfat and cream. The free fatty acids resultants by the action of lipases on milk fat endow many dairy products with their specific flavor characteristics.

A whole range of microbial lipase preparations has been developed for the cheese manufacturing industry: *Mucor meihei* (Piccnate, Gist-Brocades; Palatase M, Novo Nordisk), *A. niger* and *A. oryzae* (Palatase A, Novo Nordisk; Lipase AP, Amano; Flavor AGE, Chr. Hansen) and several others (Hasan et al., 2005).

Ester synthesis by means of lipase is an interesting alternative considering that there are many well known flavor esters in the natural aroma of fruits, traditionally obtained by extraction or by chemical synthesis (Christen and López-Munguía, 1994).

Ethyl and hexyl esters are among the important and versatile components of natural flavors and fragrances. The worldwide market for natural "green notes" is estimated to be 5–10 metric tonnes per year. Ethyl valerate, with a typical fragrance compound of green apple and hexyl acetate with a pear flavor property are in high demand and are widely used in food, cosmetic, and pharmaceutical industries. Traditionally, this kind of compounds have been isolated from natural sources or produced by chemical synthesis. Nowadays, there is a growing demand for natural flavors.

Lipase catalyzed bio-transformations are gaining importance because of their regio-, stereo-, and substrate specificity's, their milder reaction conditions and the relatively lower energy requirement. Enzymatic reactions can be efficiently accomplished by employing lipases in an adequate organic solvent, such as heptane and hexane, thereby shifting the reaction equilibrium toward esterification rather than hydrolysis. Lipases have been employed for direct esterification and trans-esterification reactions in organic solvents to produce esters of glycerol, aliphatic alcohols, and terpene alcohol's. However, the esterification of short chain fatty acids and alcohol's has not received much attention. Moreover, short-chain (C5) acids and alcohols are found to exert inhibitory effects on the enzyme. Data generation for esterification with such substrates is important because short-chain esters are vital components of many fruit flavors.

Few attempts have been made to establish the feasibility of synthesizing iso-amyl acetate by employing lipases from different sources. However, the use of high enzyme concentration and low yields has been the significant drawbacks. Also the lipase catalyzed esterification reaction has not been examined thoroughly. Reports on the use of high enzyme concentration leading to 80% yields are available. Considering the high demand and benefits, an optimized process for high yield enzymatic synthesis of iso-amyl esters, utilizing low enzyme contents, is important. The relationship between the important esterification variables (substrate and enzyme concentrations, and incubation period) and ester yield to determine optimum conditions for the synthesis of iso-amyl acetate catalyzed by Lipozyme (Krishna et al., 2000).

Esterases

Carboxyl-esterases (E.C. 3.1.1.1, carboxyl ester hydrolases) are enzymes widely distributed among all forms of life; their physiological functions have been implicated in carbon source utilization, pathogenicity, and detoxification (Ewis et al., 2004). This has a number of unique enzyme characteristics such as substrate specificity regio-specificity, and chiral selectivity. These enzymes preferably catalyze the hydrolysis of esters composed of short-chain fatty acids are preferably catalyzed by the enzymes, but they also can catalyze other reactions, like ester synthesis, and trans-esterification reactions. Notably, the interest of a broad range of industrial fields like foods, pharmaceuticals, and cosmetics has been attracted by the potential application of these enzymes for the synthesis of short-chain esters. Compounds with a great application due to their characteristic fragrance and flavor are constituted by flavor acetates from primary alcohols, which are included among the present esters.

Bacillus species has the advantages of extracellular enzyme production and lack of toxicity (most of *Bacillus* species are considered GRAS by FDA, USA). Additionally, species of the genus *Bacillus* have a history of safe use in the elaboration of traditional fermented food products, such as condiments and sauces produced from grains, legumes, and seafood. Also, safe industrial use of these bacteria and their enzymes was reported, including the production of food additives and probiotic products. Torres et al. (2009) produced and characterized a bacterial esterase, from a wild type organic solvent-tolerant *Bacillus licheniformis* S-86. One of the esterases from *B. licheniformis* S-86, called type II esterase, was previously purified in a five-step procedure. This enzyme was demonstrated to be a carboxyl-esterase specific for short-chain acyl esters, and stable and active in the presence of hydroxylic organic solvents.

The most important features about this enzyme are its proven utility for the synthesis in non-aqueous media of a valuable flavor compound with a great application in food industries. The use of type II esterase from *B. licheniformis* S-86 can be an alternative to be explored in order obtain iso-amyl acetate by "green chemistry technology" (Torres et al., 2009).

In Japan, esters are the most important flavor compounds, and the mechanisms of their formation have been investigated in great detail. Vinegar also contains many flavor components, including esters. However, little is known about how esters are produced in a particular fermentation environment in vinegar.

During vinegar production, esters are the major flavor compounds produced by *Acetobacter* sp. Kashima et al. (2000) on analyzing the relationship between ethyl acetate production and the extracellular ethanol and acetic acid concentrations found that the highest amount of ethyl acetate was produced when the molar ratio of ethanol and acetic acid was 1:1. These results indicate that the ester production by *Acetobacter* sp. is mostly catalyzed by the intracellular esterase, esterase-1, with ethanol and acetic acid used as the substrates.

Alcohol Acetyl-transferases

Esters generally have a low odor threshold: 20–30 ppm for ethyl acetate and around 1.2 ppm for iso-amyl acetate. Because the concentration of most esters formed during natural fermentations hovers around their respective threshold values, slight variations in their concentration may have dramatic effects on the beer flavor. Therefore, understanding the mechanisms of their formation to control their levels in the end product is of major industrial interest. As a consequence, the biochemical background of ester synthesis has been intensively studied.

Acetate ester formation occurs intra-cellularly through an enzyme-catalyzed reaction between acetyl-CoA and an alcohol. In the brewer's yeast *Saccharomyces cerevisiae*, two ester synthesizing enzymes have been identified: the alcohol acetyl-transferases I and II (AATase I and II; EC number: 2.3.1.84). The corresponding genes were cloned and named *ATF1* and *ATF2*, respectively. *Saccharomyces pastorianus*, *Saccharomyces carlsbergensis*, and *Saccharomyces bayanus* are other sources that can produce AATase. Furthermore, the activity of the enzymes was not limited to the acetylation of iso-amyl alcohol (resulting in the formation of the important banana flavor iso-amyl acetate), but also other alcohols, like propanol, butanol, and phenyl ethanol are esterified by *ATF1* and *ATF2*. This opens up the exciting possibility to tailor yeast's aroma production to meet consumer preferences. Not only *Saccharomyces* produced AATase, but *Kluyveromyces lactis* is commonly used in mould surface ripened cheese, and this yeast appeared to induce fruity characteristics when cultured in a cheese model medium. The alcohol acetyl-transferase orthologue of *K. lactis* is most likely responsible for part of this fruity flavor formation (Van Laere et al., 2008).

The characteristic fruity odors of wine, brandy, and other grape-derived alcoholic beverages are primarily due to a mixture of hexyl acetate, ethyl caproate (apple-like aroma), iso-amyl acetate (banana-like aroma), ethyl caprylate (apple-like aroma) and 2-phenylethyl acetate (fruity, flowery flavor with a honey note). The synthesis of acetate esters such as iso-amyl acetate and ethyl acetate in *S. cerevisiae* is ascribed to at least three acetyl-transferase activities, namely alcohol acetyl-transferase (AAT), ethanol acetyl-transferase, and iso-amyl AAT. These acetyl-transferases are sulfhydryl enzymes which react with acetyl coenzyme A (acetyl-CoA) and depending on the degree of affinity, with various higher alcohols to produce esters. It has also been shown that these enzymatic activities are strongly repressed under aerobic conditions, and by the addition of unsaturated fatty acids to a culture. The ATF1-encoded AAT activity is the best-studied acetyl-transferase activity in *S. cerevisiae*. It has been reported that the 61-kDa ATF1 gene product (*ATF1*p) is located within the yeast's cellular vacuomes and plays a major role in the production of iso-amyl acetate and to a lesser extent ethyl acetate during beer fermentation. The AAT in wine and brandy composition, has cloned, characterized, and mapped the *ATF1* gene from a widely using commercial wine yeast strain, VIN13, that was to over express the *ATF1* gene during fermentation to determine its effect on the yeast metabolism, acetate ester formation, and flavor profiles of Chenin blanc wines and distillates from Colombar base wines. Ultimately, lead to the development of a variety of wine yeast strains for the improvement of the flavor profiles of different types and styles of wines and distillates, especially of those products deficient in aroma and lacking a long, fruity shelf life (Lilly et al., 2000).

Acyl-CoA

According to Aguedo et al. (2004), acyl-CoA is involved to generation of aroma compounds in this case, γ-decalactone through ß-oxidation is the classical biochemical pathway involved in fatty acid degradation, it acts on an acyl-CoA and consists of a four-steps reaction sequence, yielding an acyl-CoA which has two carbons less and an acetyl-CoA. This can lead to a variety of volatile compounds (Fig. 1.1).

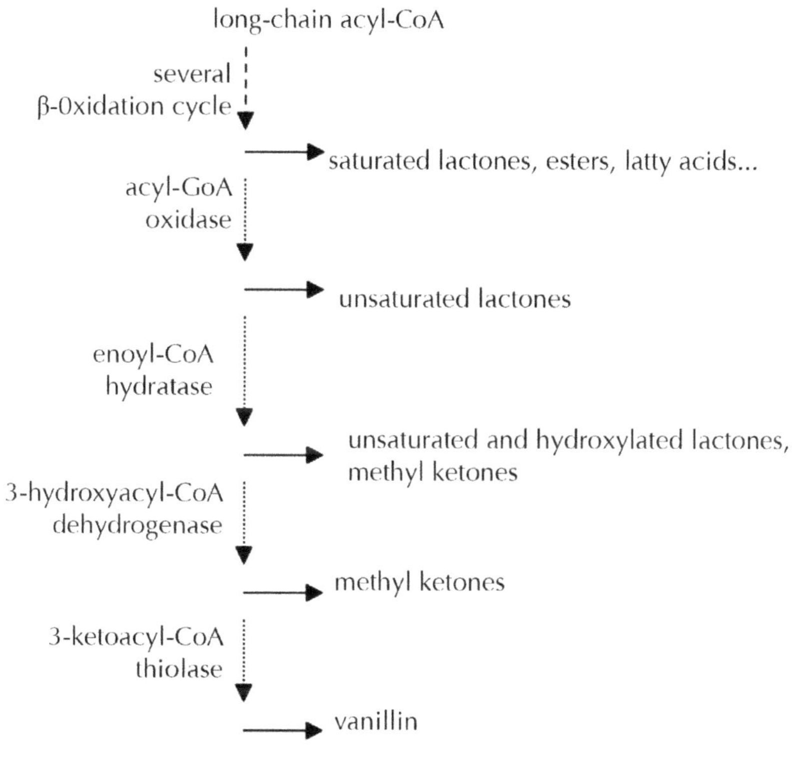

Figure 1.1. β-Oxidation cycle and accumulation of aroma compounds.

ENZYMES FOR DAIRY INDUSTRY

Since the origin of cheese production, at least 4000 years, the manufacturing process has been adapted to current technologies. Rennet, also called rennin, is a mixture of chymosin and pepsin obtained from the fourth stomach of suckling calves. Over the years there was an imbalance between the production of cheese and availability of animal rennet. Microbial enzymes have therefore been introduced as alternative. Chymosin is an acid protease that cleaves specifically the peptide bond between the phenylalanine and methionine, from N-terminus of the protein content. Submerged fermentation with fungi such as *Rhizomucor miehei* produces microbial rennets with similar properties to those of chymosin. In practice, microbial enzymes have lower milk coagulating activity and proteolytic activity than pure calf rennet (Soccol et al., 2005).

Proteases are used in order to influence the organoleptic properties (aroma and flavor) of cheese, which in turn depend on the relative amounts of amino acids and fatty acids throughout the product. Proteases are also used for accelerating cheese ripening, for modifying the functional properties of cheese and for modifying milk proteins to reduce the allergic properties of dairy products.

Other components of milk are allergens such as lactose. The enzyme lactase (β-galactosidase) is used to hydrolyze lactose in order to increase digestibility, or to improve solubility or sweetness of various dairy products. The main sources of lactase are the yeasts *Kluyveromyces fragilis, K. lactis* and the fungi *A. niger* and *A. oryzae.*

Some of the applications of the hydrolysis of lactose in dairy products are: (a) lactose removal from milk powder or liquid allows it to be ingested by disabled lactase. Moreover, in the case of flavored milk increases the rate of sweetness and enhances the flavor of this product; (b) reduce the time of fermentation of fermented dairy products; (c) prevent the crystallization of lactose in ice cream and condensed milk. Remember that lactose concentration in the order of 12% compared to the total milk solids (e.g., skimmed), crystallizes easily. When 50% reduction in the level of lactose in milk, it can be used in larger amounts in formulations, which reduces the amount of stabilizer to be added, since this function can be performed by proteins own milk. Furthermore, since the lactose hydrolyzed milk is the number of free molecules increased, which contributes to the freezing point depression solution, when it is used in formulations of creams, and d) The removal of lactose whey allows the manufacture of syrups with 70–75% of solid and can be used in food formulations (Soccol et al., 2005).

ENZYMES FOR BAKING INDUSTRY

Wheat flour is one of the main ingredients of bakery products. Starch and proteins are the major constituents of wheat flour. Alpha amylase is a special type of amylase which modifies starch during baking to give a significant anti-staling effect.

Besides the traditional enzymes such malt and α-amylases, the advances in biotechnology have made interesting enzymes available for bread making. Hemi-cellulases, xylanases, lipases, and oxidases can directly or indirectly improve the strength of the gluten network, and so improve the quality of finished bread (Soccol et al., 2005).

Within the baking industry, phospho-lipases can be supplied as effective substitute or supplement traditional emulsifiers, as the enzymes degrade polar wheat lipids to produce emulsifying lipids *in situ*. Lipases from *A. niger*, *R. oryzae*, *C. cylindracea* are used in bakery products to increase the volume in bread and bakery (Hasan et al., 2005).

Also, efforts are currently devoted toward the further understanding of bread staling and the mechanisms behind the enzymatic prevention of staling when using α-amylases and xylanases. Studies have confirmed previous findings showing that water-binding capacity and retention in the starch and hemi-cellulose fractions of the bread, being the substrates of α-amylases and xylanases, respectively, to be critical for maintaining softness and elasticity. This amylase is probably capable of degrading amylo-pectin to a degree that prevents re-crystallization after gelatinization, without completely degrading the amylo-pectin network which provides the bread with elasticity (Kirk et al., 2002).

Addition of enzymes to sourdough sponges can also enhance bread volatile compounds. According to Rehman et al. (2006), lipid oxidations through addition of enzymes in the form of active soya flour increase concentrations of hexanal, 1-hexanol, 1-penten-3-ol, 1-pentanol and 2-heptanone, and 2-heptenal and 1-octen-3-ol, only in breads containing soya flour. Addition of lipase, endo-xylanase, and α-amylases enhanced acetic acid production by *Lb. hilgardii* 51B. Textural analyses suggest that such sourdoughs with single enzyme have greater stability and crumb softening than breads produced with only exogenous enzymes (Rehman et al., 2006).

The uses of enzymes enable bakeries to extend shelf-life of breads, enhance, and control non-enzymatic browning, increase loaf volume and improve crumb structure. There are many industrial manufactures of a wide and continuously evolving range of bakery enzymes to fulfill the needs of the bakery trades.

OTHER APPLICATIONS

Besides the advances observed, a few new applications of enzymes within the food industry could be mentioned.

Enzymes are used also in drinks and fruit juice manufacturing. Addition of pectinase, xylanase, and cellulase improve the liberation of the juice from the pulp. Pectinases and amylases are used in juice clarification. Almost all the commercial preparations of pectinases are produced from fungal sources. *Aspergillus niger* is the most commonly used fungal species for industrial production of enzymes (Jayani et al., 2005). The use of lactase for clarification of juice (lactases catalyze the cross-linking of poly-phenols, resulting in an easy removal of poly-phenols by filtration) and for flavor enhancement in beer are recently established applications within the beverage industry. Brewing is an enzymatic process. Malting is a process, which increases the enzyme levels in the grain. In the mashing process the enzymes are liberated, and they hydrolyze the starch into soluble fermentable sugars like maltose, which is a glucose disaccharide (Ghorai et al., 2009).

In tea processing, the quality of black tea is dependent great extent on the dehydration, mechanical breaking, and enzymatic fermentation to which tea shoots are subjected. During manufacture of black tea, enzymatic breakdown of membrane lipids initiate the formation of volatile products with characteristic flavor properties emphasize the importance of lipids in flavor development, and lipase produced by *Rhizomucor miehei* enhanced the level of polyunsaturated fatty acids observed by reduction in the total lipid content (Hasan et al., 2005).

Trans-glutaminase has been applied as a texturing agent in the processing of sausages, noodles, and yoghurt, where cross-linking of proteins provides improved visco-elastic properties of the products. Their wider use is currently limited due to availability of the enzyme in industrial scale. At present only the trans-glutaminase from

Streptoverticillium species is commercially available at a reasonable scale, and work is ongoing to increase the availability of the enzyme by recombinant production in *Escherichia coli* (Kirk et al., 2002).

CONCLUSION

The field of industrial enzymes has received great efforts in research and development, resulting in both the development of a number of new products, and in improvement in the process and performance of several existing products. Microbial enzymes, looking at the increased awareness of environment and cost issues, is gaining ground rapidly due to the various advantages of biotechnology, over traditional technologies.

New applications of enzymes within food industry will depend on the functional understanding of different enzyme classes. Furthermore, the scientific advances in genome research and their exploitation via biotechnology is leading to a technology driven revolution that will have advantages for the consumer, and food industry alike.

KEYWORDS

- **Enzymes**
- **Glycosidases**
- **Lipases**
- **Microorganisms**
- **Proteases**

Chapter 2

Efficient Extraction and Utilization of Litchi (*Litchi sinensis* Sonn.) Fruit

K. Nagendra Prasad, Amin Ismail, Muhammad Ashraf, and Yueming Jiang

INTRODUCTION

Litchi (*Litchi chinensis* Sonn.) is a tropical to subtropical crop that originated in South-East Asia. Litchi fruit are prized on the world market for their flavor, semi-translucent white aril and attractive red skin. Litchi is now grown commercially in many countries and its production in Australia, China, Israel, South Africa, and Thailand has expanded markedly in recent years. Increased production has made significant contributions to economic development in these countries, especially those in South-East Asia. Litchi has been used since ancient age for the traditional Chinese medicine formulation, serving as relief of neural pain and swelling. Litchi fruit pericarp (LFP) tissues account for approximately 15% by weight of the whole fresh fruit and are comprised of significant amount of flavonoids. Furthermore, the extract from LFP tissues exhibits excellent antioxidant ability and good anticancer activity. The book chapter overviews the recent advances in the extraction, biological activities, and utilization of the bioactive compounds present in LFP. Some novel pharmacological potential of the litchi fruit is also discussed, which could be helpful for full utilization of this fruit in the future.

Litchi (*Litchi chinensis* Sonn.) is most commonly grown in tropical to subtropical regions of the world. It belongs to *Sapindaceae* family and is native to China, India, Vietnam, Indonesia, and the Philippines and now it is grown in other parts of the world. The fruit is of high commercial value because of its white translucent pulp and bright red color pericarp, with sweet taste and pleasant aroma (Jiang et al., 2006; Prasad et al., 2009a; Zhang et al., 2004a). The fruit is rich in nutrition with carbohydrates, protein, fiber, fat, vitamin C, amino acids, and minerals like phosphorus, calcium, magnesium, potassium, sodium, copper, iron, zinc, manganese, and boron (Hall et al., 1980; Mahattanatawee et al., 2006; Sivakumar et al., 2008; Wall, 2006; Yang et al., 2010). The fruit seeds are used as traditional Chinese medicine to cure neural pain, swelling, and cough (Li, 2008).

The LFP tissues account for approximately 15% by weight of the whole fresh fruit and are comprised of significant amount of flavonoids (Li and Jiang, 2007). Therefore, LFP tissues may be considered an important source of dietary flavonoids. Furthermore, the fruit pericarp is rich in anthocyanins such as cyanidin-3-rutinoside, cyanidin-3-glucoside, malvidin-3-glucoside, and pelargonidin (Lee and Wicker, 1990; Prasad and Jha, 1978; Zhang et al., 2004a), phenolic compounds like gallic acid, catechin, epicatechin, gallocatechin, quercetin, rutin, procyanidin B_1, procyanidin B_2, epicatechin-3-gallat, and tannins (Li and Jiang, 2007; Mahattanatawee et al., 2006;

Prasad et al., 2009b; Roux et al., 1998; Sarni-Manchado et al., 2000; Zhang, 2000; Zhao et al., 2006), polysaccharides (Kong et al., 2010; Yang et al., 2006) and some volatile compounds (Sivakumar et al., 2008). Litchi fruit has been used since ancient age for the traditional Chinese medicine formulation, serving as relief of neural pain and swelling. In recent years, the extract from LFP tissues has exhibited excellent antioxidant ability and good anticancer activity (Li and Jiang, 2007; Zhao et al., 2007). Thus, LFP tissues can be used as a readily accessible source of natural antioxidants and/or a possible supplement in the food or pharmaceutical industries. To better utilize LFP tissues, the book chapter overviews the recent advances in the extraction, biological activities and utilization of bioactive compounds. In addition, some novel pharmacological potential of LFP is discussed, which could help a full utilization of this fruit in the future.

EXTRACTION

Extraction is the most important step in isolating different types of bioactive compounds from LFP tissues. The ideal extraction method should be quantitative, non-destructive, and time-saving. There are different types of extraction methods, but a variety of methods are in vogue for the extraction of the bioactive compounds from LFP. Each method has its own advantage and disadvantage. Recently, usage of green technology like high pressure or ultrasonic extraction as an environment-friendly technology has been reported for extraction of bioactive compounds from LFP.

Conventional Extraction

It is the oldest and traditional way of extracting bioactive compounds from plant tissues, using soxhlet extractor, magnetic stirrer, boiling, maceration, grinding, and heat reflux. These extraction techniques depend largely on the type of solvents, energy input, and agitation to improve the chemical solubility and efficiency of mass transfer. Usually, these methods require long extraction time with low extraction yield but high energy consumption (Xiao, 2003). These methods are still popular and in use because they are easy to perform and cheap to operate.

Roux et al. (1998) reported that proanthocyanidins can be extracted from the dried LFP by maceration for 30 min with acidified methanol. Sarni-Manchado (2000) extracted phenolic compounds from LFP at 4°C for 20 min using acidified methanol. Anthocyanins can be extracted well from LFP tissues using 90% acidified methanol for 30 min at 4°C (Zhang et al., 2000). Frozen LFP was also extracted to obtain anthocyanins overnight at 25°C with 0.5 M HCl (Zhang et al., 2004a). Yang et al. (2006) extracted polysaccharides using water for two hours from LFP. In addition, anthocyanins were isolated from LFP using 0.5 M HCl for 12 hr (Duan et al., 2007a) while phenolic compounds were also extracted using 65% ethanol at 4°C for 30 min (Duan et al., 2007b). Liu et al. (2007) extracted procyanidins after four hours of extraction with acetone from LFP. The conventional way of extracting flavonoids from LFP was investigated further using 85% acidified ethanol as an extracting solvent (Zhou et al., 2006, 2007). Furthermore, the extraction conditions to obtain higher extraction yield from LFP were optimized (Prasad et al., 2009a). The conventional way of extracting flavonoids from LFP was investigated also using 85% acidified ethanol for 24 hr (Prasad et al., 2009b). The extraction of phenolic compounds from litchi seeds using different solvents like ethanol, 50% ethanol, methanol, 50% methanol, and water was carried out also for 12 hr using rotary shaker (Prasad et al., 2009C). Recently, Kong et al. (2010) have extracted polysaccharides from LFP using 80% ethanol for four hours at 80°C. It is noted that different conventional extraction methods have been adopted using various solvents to extract anthocyanins, flavonoids, polyphenolic compounds, polysaccharides, and so forth, due to a large number of litchi cultivators available and they require to be optimized further.

Ultrasonic Extraction

Ultrasonic extraction (UE) is the most industrially-oriented method to improve mass transfer. The cavitation force occurring in this method can eliminate the bubbles and generate localized pressure to rupture plant tissues. As a result, the release of intracellular substances into the solvent is improved (Knorr et al., 2002). Usage of ultrasonication was shown to attack the integrity of plant cellular walls, resulting in the release of the extract into the solvent in a shorter time at lower temperature (Mason and Zhao, 1994). The UE of oils from chickpea (Lou et al., 2010), saponins from ginseng (Engelberth et al., 2010), and polysaccharides (Yang et al., 2008) and corilagin from longan fruit pericarp (Prasad et al., 2010) was reported. Using the optimum conditions, the application of UE to obtain a higher extraction yield from LFP has been demonstrated using 85% acidified ethanol (Prasad et al., 2009a). UE to extract flavonoids from LFP was conducted using 85% acidified ethanol as the extraction solvent at room temperature for 24 hr (Prasad et al., 2009b). Rao et al. (2009) optimized UE to isolate coenzyme Q10 from LFP tissues. Thus, the efficacy of ultrasonication to extract types and yields of bioactive compounds from litchi fruit is worth further investigation.

High Pressure Extraction

High pressure extraction (HPE) method, which works under very high pressure ranging from 100 to 1000 MPa, has been recognized as an environment-friendly technology by the Food and Drug Administration (FDA), and is being extensively applied in pharmaceutical, metallurgical, and food industries (US, FDA, 2000; Zhang et al., 2004b). One of the most promising advantages of HPE is that pressure acts instantaneously and uniformly through a food mass independent of its shape, size, or composition (Ahmed and Ramaswamy, 2006). High pressure may lead to various structural changes in foods, like damage and deformation of cell and its membranes as well as denaturation of proteins and enzymes (Corrales et al., 2008). High pressure can effectively enhance the efficiency of mass transfer and improve solvent permeability in the cells as well as diffusion of secondary metabolites, but had no negative effect on the structure and activity of bioactive compounds (Ahmed and Ramaswamy, 2006; Dornenburg and Knorr, 1993). Thus, HPE exhibits many advantages for extraction of natural products. Various researchers have successfully used the HPE technique for the extraction of corilagin from longan (Prasad et al., 2010) and anthocyanins from grapes (Corrales et al., 2008). Optimization to obtain the high extraction yield from LFP was determined using different high pressures (200-500 MPa), high pressure time

(2.5–30 min) and high pressure temperature (30–90°C). Table 2.1 shows a higher extraction yield by application of high pressure for 2.5 min than the conventional extraction for 24 hr or the ultrasonic extraction for 30 min. The results exhibit that HPE is more effective as compared to the other methods in extracting bioactive compounds from LFP (Prasad et al., 2009a). Additionally, it should be noted that flavonoids obtained by HPE is almost 100 times higher compared to the conventional extraction (Table 2.2).

Table 2.1. Comparison of extraction yield and extraction time by conventional, ultrasonic UE and HPE methods (Modified from Prasad et al. (2009a).

Extraction method	Extraction time	Extraction yield (%)
Conventional	24 hr	19.9 ± 1.12
Ultrasonic	30 min	23 ± 0.5
High Pressure (500 MPa)	2.5 min	29.3 ± 0.19

Table 2.2. Quantification of individual flavonoids from litchi fruit pericarp tissues using different extraction methods (Modified from Prasad et al. (2009a).

Flavonoids (mg/g DW)	Extraction methods		
	Conventional	Ultrasonics	High pressure (400 MPa)
Epicatechin	0.0414 ± 0.001	0.16 ± 0.04	0.348 ± 0.06
Epicatechin gallate	0.0121 ± 0.003	0.06 ± 0.01	0.2527 ± 0.04
Catechin	0.0002 ± 0.0	0.0020 ± 0.0005	0.0160 ± 0.07
Procyanidin B_2	0.0175 ± 0.0003	0.0731 ± 0.0011	0.1346 ± 0.03
Total content	0.0712 ± 0.004	0.2951 ± 0.051	0.7513 ± 0.2

PHARMACOLOGICAL PROPERTIES

Litchi is an important commercial fruit in Asian countries. Since centuries, litchi fruit has been used as the traditional medicine to relieve neural pain, cough, and swelling in China (Li, 2008). In recent years, other pharmacological activities such as antioxidant activity and anticancer properties of litchi fruit have been determined. Anthocyanins, phenolic compounds, flavonoids, polysaccharides, and tannins present in litchi fruit can be responsible for these bioactivities.

Antioxidant Activity

In recent years, the exploitation of natural antioxidants particularly from plant origins instead of synthetic antioxidants has been recommended. A variety of natural antioxidants have been identified in different types of fruits and vegetables, with the most common antioxidants being phenolic compounds. The phenolic compounds comprise phenolic acids, flavonoids, anthocyanins and hydrocinnamic acid derivatives (Halliwell, 2007; Li and Jiang, 2007).

Antioxidant activities of peel, pulp, and seeds of litchi fruit were determined by ferric reducing antioxidant power (FRAP) method, with their values being 0.59, 2.8, and 22.3 mmol/100 g, respectively (Guo et al., 2003). Antioxidant activities of litchi puree were determined also by oxygen radical absorbance capacity (ORAC) and 1, 1-diphenyl-2-picrylhydrazyl (DPPH) methods, with a good correlation of the antioxidant activities with total phenolic contents (Mahattanatawee et al., 2006). Yang et al. (2006) determined the antioxidant activities of polysaccharides from LFP and found their antioxidant activities were concentration-dependent. The polysaccharides exhibited excellent antioxidant activities by DPPH, hydroxyl, and superoxide radical scavenging methods (Table 2.3).

Table 2.3. Radical-scavenging activities of the different polysaccharides extracted and purified from LFP tissues (Modified from Yang et al. (2006).

Sample	Concentration (μg/mL)	DPPH radical scavenging activity (%)	Hydroxyl radical scavenging activity (%)	Superoxide radical scavenging activity (%)
CP	50	17.6 ± 0.3	21.4 ± 0.4	22.7 ± 0.5
	100	19.9 ± 0.4	25.6 ± 0.7	27.6 ± 0.6
F1 fraction	50	38.4 ± 1.1	42.7 ± 1.3	43.2 ± 1.1
	100	48.5 ± 1.4	47.9 ± 1.2	47.6 ± 1.2
F01 fraction	50	40.6 ± 0.9	46.8 ± 1.4	50.7 ± 1.6
	100	54.1 ± 1.5	53.1 ± 1.8	62.5 ± 2.0
BHA	50	68.3 ± 1.1	84.2 ± 1.4	78.9 ± 1.7
	100	76.6 ± 1.2	90.7 ± 1.7	85.7 ± 1.6

CP: Crude polysaccharides; F1 fraction: polysaccharides purified partially by a DEAE sepharose chromatography; F01 fraction: polysaccharides purified further by a Sephadex G-50 gel-permeation chromatography; and BHA: Butylated hydroxyl anisole.

Zhao et al. (2006) identified three flavonoids from LFP, namely epicatechin, proanthocyanidin B_2 and proanthocyanidin B_4, and then determined their antioxidant capabilities by DPPH radical, superoxide anion and hydroxyl radical scavenging activities. All the flavonoids exhibited high antioxidant activities. Duan et al. (2007a; 2007b) found that anthocyanins derived from LFP tissues effectively suppressed linoleic acid oxidation and showed a dose-dependent free-radical scavenging activity against DPPH and hydroxyl radicals, and superoxide anion. Two polyphenol oxidase substrates namely epicatechin and procyanidin A isolated from LFP exhibited also strong antioxidant activities determined by reducing power, DPPH radical, hydroxyl radical and superoxide anion scavenging methods (Sun et al., 2010). Liu et al. (2007) reported that oligomeric procyanidins, A_2 and trimeric procyanidins isolated from LFP had a high scavenging activity against hydroxyl radical, with the corresponding IC_{50} values of 2.60, 1.75, and 1.65 μg/mL (Liu et al., 2007). Ruenroengklin et al. (2008) evaluated the antioxidant activities of anthocyanins isolated from LFP at various temperatures and pH values and then concluded that high antioxidant activity could be achieved at 45–60°C and pH 3–4. In addition, Khan et al. (2009) studied the radical scavenging activities of ethylacetate, butanol, methanol, and water extracts of pulp, stem and leaf of litchi. All the extracts showed a good antioxidant activity which can be comparable to synthetic antioxidants. Prasad et al. (2009c) investigated antioxidant activities of litchi seeds using different solvents and found that all the extracts showed excellent antioxidant activities by DPPH radical scavenging method (Fig. 2.1), lipid peroxidation (Fig. 2.2) and reducing power (Fig. 2.3). The compounds responsible for the antioxidant activity were determined to be gallic acid, gallocatechi, epicatechin, procyanidin B_2, and epicatechin-3-gallate. However, the comparison of antioxidant activities by DPPH radical and superoxide anion scavenging methods of the LFP extracts obtained from the conventional, ultrasonic, and HPEs (400 MPa) made by Prasad et al. (2009b) exhibited no significant differences in the antioxidant activities among all these extracts tested (Fig. 2.4).

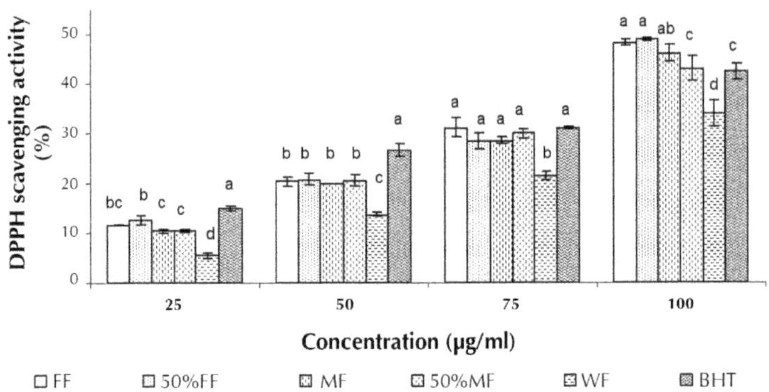

Figure 2.1. DPPH radical scavenging activity of ethanol extract (EE), 50% ethanol extract (50% EE), methanol extract (ME), 50% methanol extract (50% ME) and water extract (WE) from litchi seeds. For each treatment, the means within the same set followed by different letters were significantly different at the level of $P < 0.05$ (Modified from Prasad et al. (2009c).

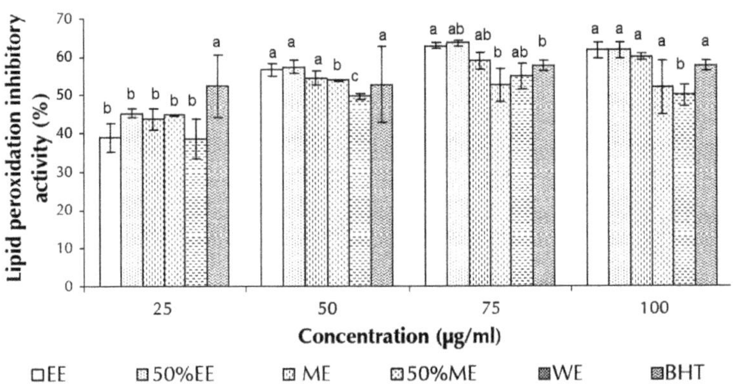

Figure 2.2. Lipid peroxidation inhibitory activity of ethanol extract (EE), 50% ethanol extract (50% EE), methanol extract (ME), 50% methanol extract (50% ME), and water extract (WE) from litchi seeds. For each treatment, the means within the same set followed by different letters were significantly different at the level of $P < 0.05$ (Modified from Prasad et al. (2009c).

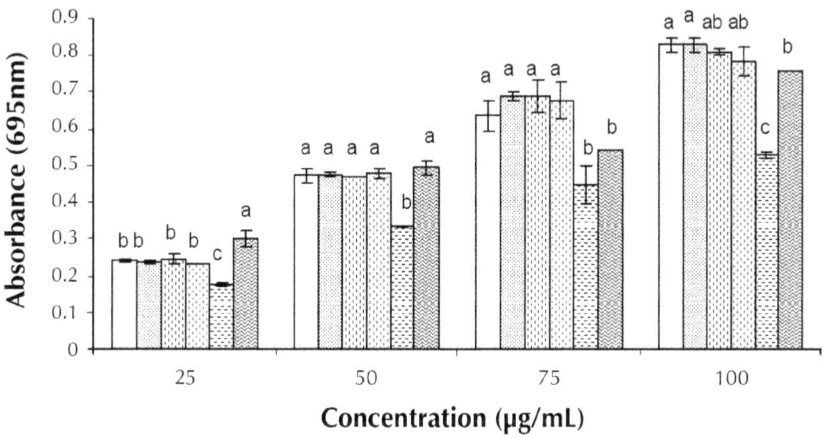

Figure 2.3. Total antioxidant activity using phospomolybdenum method of ethanol extract (EE), 50 % ethanol extract (50% EE), methanol extract (ME), 50% methanol extract (50% ME) and water extract (WE) from litchi seeds. Higher absorbance value indicates higher antioxidant activity. For each treatment, the means within the same set followed by different letters were significantly different at the level of $P < 0.05$ (Modified from Prasad et al. (2009c).

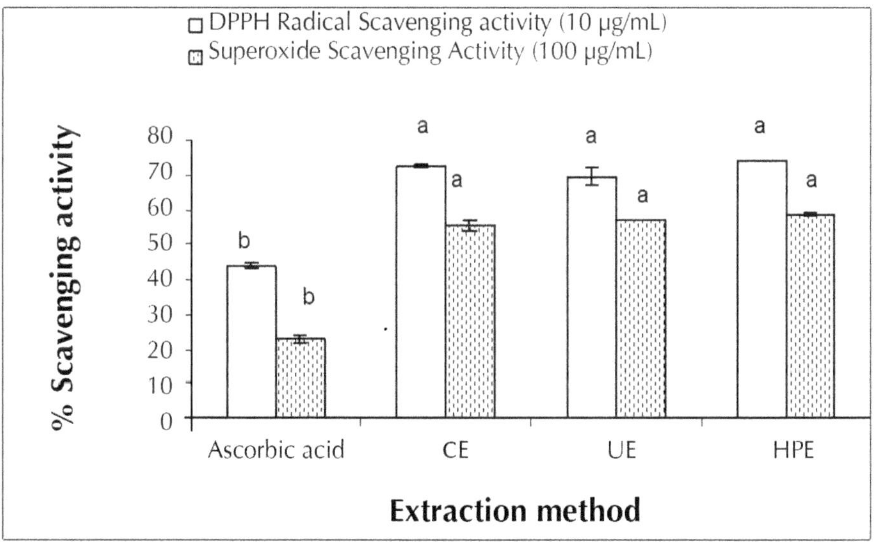

Figure 2.4. Comparison of DPPH radical scavenging activity and superoxide scavenging activity from litchi fruit pericarp using HPE, UE and CE. For each treatment means in a row followed by different letters are significantly different at $P < 0.05$ (Modified from Prasad et al. (2009a).

Anticancer Activity

Considerable attention is being paid to explore potential anticancer components from medicinal plants. Components like flavonoids and polyphenols commonly occurring in medicinal plants have been shown to reduce various types of experimental carcinogenesis such as cancer (Chen et al., 2008; Zhao et al., 2007). Epicatechin and proanthocyanidin B_2 obtained from LFP showed a strong anticancer activity against human breast cancer cell line, with IC_{50} values of 102 and 99 μg/mL, respectively, and was found to be as potent as paclitaxel (Zhao et al., 2007). A promising anticancer activity of water extract from LFP against liver cancer (hepatocellular carcinoma) using *in vitro* and *in vivo* systems have been shown by Wang et al. (2006a). The hepatoma tumor cells significantly decreased with increasing concentration of the water extract used. The tumor inhibitory rates of the water extract of LFP at 0.15, 0.3, and 0.6 g/kg/day were 17, 30, and 44%, respectively. The experimental findings showed the substantial dose-dependent suppression of tumor growth in the LFP extract-treated mice and then suggested that the water extract from LFP had a strong antiproliferation effect by causing apoptotic cell death, suppressing DNA synthesis, and arresting G0/G1 phase of cancer cells. Subsequently, the anticancer activity of the water extract of LFP against human breast cancer using *in vitro* and *in vivo* experiments could be ascribed partially to its DNA damage effect, proliferating inhibition and apoptosis induction of cancer cells through expression of a number of genes involved in cell cycle regulation and cell proliferation (Wang et al., 2006b). The extract from litchi seeds at 100 μg/mL possessed high anticancer activities against SCG-7901, A-549 and HepG-2 cancer cell lines, with relative inhibitory activities of 68.2±1.9, 45.9±3.2 and 39.57±0.95%, which were comparable to cisplatin having inhibitory values of 63.9±4, 49.57±1.3, and 55.37±2.1%, respectively (Yang et al., 2010). Thus, the constituents from the LFP extract might be considered as a powerful novel antitumor agent but the anticancer efficacy of flavanols and anthocyanins present in LFP tissues requires to be documented further.

UTILIZATION AND PROSPECTS

Litchi fruit is gaining popularity throughout the world because of its taste, color, and aroma. The prospect of increasing production of litchi and its relatively short post-harvest life as a fresh commodity, raises expectations for increased processing and utilization opportunities for this crop. The usage of litchi fruit in China dates back to 2000 years ago. Additionally, litchi seeds are also used in the traditional Chinese medicine. LFP is a rich source of anthocyanins, flavonoids, phenolic compounds, tannins, and polysaccharides, with many health benefits like antioxidants, anticancer, and so on. With the recent advancement in green technologies like HPE and UE, bioactive compounds from LFP can be extracted efficiently. Although litchi fruit is commonly eaten fresh and also dried for preservation, some litchi-derived food is produced in the market, like canned litchi, jam, syrup, sauce, tea, gel, wine, juice, ice-cream, and yoghurt (Husain, 2009; Li, 2008). In addition, the bioactive compounds from LFP as natural antioxidants instead of the synthetic antioxidants such as butylated hydroxyanisole (BHA) and butylated hydroxytoluene (BHT), which can increase the shelf-life and reduce the deterioration of food products during processing and storage is being used commercially. Furthermore, some functional foods from LFP in relation to their beneficial effects against gastralgia, tumors, and enlargements of the glands of consumers are being developed.

ACKNOWLEDGMENTS

Financial support from the CAS/SAFEA International Partnership Program for Creative Research Teams, the National Natural Science Foundation of China (Grant No. 30425040), and the Natural Science Foundation of Guangdong Province (Grant No. 06200670) is greatly appreciated.

KEYWORDS

- **Dietary flavonoids**
- **Polyphenolic**
- **Polysaccharides**
- ***Sp.***
- **Triacyl-glycerohydrolases**

Chapter 3

Extraction, Composition, and Functional Properties of Pectin from Chickpea Husk

Vania Urias–Orona, Agustín Rascyn–Chu, Jaime Lizardi, Elizabeth Carvajal–Millán, Alfonso Gardea, and Alma Rosa Islas

INTRODUCTION

Pectin is one of the main structural components of plant cell walls. Pectin forms gels under certain circumstances, the gelling mechanism is highly dependent on the degree of methoxylation (DM). Pectin conventionally is divided into high methoxy (HM) pectin with DM>50% and low methoxy (LM) pectin with DM<50%. HM pectin forms gels in the presence of high sugar concentration, usually sucrose or fructose and low pH. Conversely, LM pectin forms gels by interaction of carboxyl group ionized and divalent ions by the "egg box" mechanism. The viscoelastic properties of pectin are the base of its broad use as gelling agent and stabilizer in food products. Although most plant tissues contain pectin, citrus peel, and apple are the major sources of pectic substances around the world. Although most chickpeas are produced for human consumption, low quality grains provide the livestock industry with an alternative protein and energy feedstuff. A large amount of by-products is produced during chickpea processing in regions where this is a major food legume. The majority of chickpea processing wastes include chickpea husk, which is used for animal nutrition. In this chapter, chickpea husk pectin extraction, composition, and functional properties were investigated. Pectin presented a 67% (w/w) of galacturonic acid, an intrinsic viscosity of 374 ml/g and a viscosimetric molecular weight of 110 kDa. Fourier transform infrared (FTIR) spectroscopy spectrum of pectin indicates a degree of esterification of 10%. For 2% (w/v) calcium induced pectin gels, the values of the storage (G¢) and loss (G¢¢) modulus and the gel set time (tg) were 57 and 10 Pa, and 3.5 min, respectively. After 48 hr storage at 25°C, pectin gels showed no significant difference on the texture profile analysis, except for hardness (21% loss). Pectin whippability, foam stability, emulsion stability, and reduced viscosity were maximum at pH 9 (124, 95, and 96% and 61 ml/g, respectively). The results attained suggest the use of this pectin as potential texturing agent for the food industry.

Pectins are natural hydrocolloids found in higher plants that are widely used as gelling agents, stabilizers, and emulsifiers in the food industry (May, 1990). Pectin is an heteropolysaccharide consisting of homogalacturonan ("smooth" regions), composed of an α-(1→4)D-GalA*p* chain and rhamnogalacturonan ("hairy" regions) consisting of the repeating disaccharide GalA*p*-a-(1→2)-Rha*p*-a-(1→4)-GalA *p*a-(1→2)-Rha*p*). Arabinose and galactose side chains can also be attached creating heteropolysaccharide complexes of rhamnogalacturonan with arabinans, galactans, and arabinogalactans (Willats et al., 2006). Pectin forms gels under certain circumstances but the gelling mechanism is highly dependent on the DM. Conventionally, pectin is divided into HM pectin with DM>50% and LM pectin with DM<50%. HM pectin forms gels in the presence of high sugar concentration, usually sucrose or fructose and low pH; whereas LM pectin does, in the presence of divalent ions for example calcium (Ström et al., 2007). Pectin is widely used in jams and jellies, fruit preparations, fruit drink concentrates, fruit juices, desserts, and fermented dairy products (May, 1990). Although most plant tissues contain pectin, commercial production around the world is based almost entirely on citrus and apple peel (Rascón-Chu et al., 2009). Mexico is one of the major producers of chickpea (*Cicer arietinum* L.), which is mainly exported as this grain does not represent an important constituent of the Mexican diet. In regions where chickpea is a major food legume (Southern Europe, North Africa, India, and Middle East countries) a large amount of by-products are generated. The majority of chickpea processing wastes include chickpea husk (Christodoulou et al., 2005; Maheri-Sis et al., 2008), which is used for animal nutrition. A previous research indicated that chickpea could be a source of pectin (Aisa et al., 2006). Nevertheless, the characterization and functional properties of this chickpea husk pectin have not been yet reported elsewhere. In this chapter the extraction, composition, and functional properties of pectin from chickpea husk are presented.

EXPERIMENT

Materials

Chickpea (Mocorito-88 variety) sample was kindly provided by the National Institute for Investigation in Forestry, Agriculture and Animal Production in Mexico (INIFAP-CEVACU). All chemical products were purchased from Sigma Chemical Co. (St Louis, MO, USA).

Pectin Extraction

Chickpea seeds were first heated (1 kg seeds/2 l water) for 15 min at 50°C. The husks were then manually separated, dried at 40°C overnight and milled down to 0.84 mm particle size. Milled husks were dispersed in phosphate buffer (100 g/600 ml) and treated enzymatically for starch and protein degradation, using α-amylase solution (Termamyl®120, pH 7, 100°C, 30 min, 75 U/g sample), amyloglucosidase, (two hours, 60°C, pH 5.5, 240 U/g sample, 80 rpm) and pronase (pH 7 at 25°C to 18 hr, followed for 100°C, 10 min, 0.4 U/g sample). Pectin extraction was performed twice under acid conditions, using 0.05 N HCl (1:6) at 80°C for one hour and 80 rpm and both supernatants were collected. The extract was centrifuged at 12,040 g for 10 min and the pH adjusted to 3.5. The extract was dispersed into three volumes of 96% ethanol during one hour at 4°C in order to precipitate pectin, which was then collected by filtration through 4 µm (Whatman) and freeze-dried.

METHODS

Sugar Composition

Sugar composition was determined according to Carvajal-Millan et al. (2007) after pectin hydrolysis with 2 N trifluoroacetic acid at 120°C for two hours. The reaction was stopped on ice, the extract was evaporated under air at 40°C and rinsed twice with 200 µl of water and resuspended in 500 µl of water. All samples were filtered through 0.45 µm (Whatman) and analyzed by high performance liquid chromatography (HPLC) using a Supelcogel Pb column (300 × 7.8 mm; Supelco, Inc., Bellefont, PA, USA) eluted with 5 mM H_2SO_4 (filtered 0.2 µm, Whatman), at 0.6 ml/min and 50°C. A refractive index detector Star 9040 (Varian, St. Helens, Australia) and a Star Chromatography Workstation system control version 5.50 were used. The internal standard was inositol.

Degree of Esterification (DE)

Pectin degree of esterification was determined by FTIR spectroscopy (Nicolet Instrument Corp. Madison, WI, USA) as described by Gnanasambandan et al. (2000).

Intrinsic Viscosity

Specific viscosity, η_{sp} was measured by registering pectin solutions flow time in an Ubbelohde capillary viscometer at 25 ± 0.1°C, immersed in a temperature controlled bath. Pectin solutions were prepared at different concentrations, dissolving dried pectin in an aqueous solution containing 0.1 N NaCl at pH 7 for 18 hr with stirring at room temperature. Pectin solutions and solvent were filtered using 0.45 µm membrane filters before the viscosity measurements. The intrinsic viscosity ($[\eta]$) was estimated from relative viscosity measurements, η_{rel}, of pectin solutions by extrapolation of Kraemer and Mead and Fouss curves to "zero" concentration (Kraemer, 1938; Mead and Fouss, 1942). NaCl was used in order to prevent pectin aggregation.

Viscosimetric Molecular Weight

The viscosimetric molecular weight (Mv) was calculated from the Mark–Houwink relationship, $Mv = ([\eta]/k)^{1/\alpha}$, where the constants k and α are 0.0436 and 0.78, respectively.

Pectin Gelation

Pectin solution was prepared at 2% (w/v) dissolving dried pectin in water at pH 7 for 18 hr with stirring at room temperature. The gelling pectin–calcium mixture was prepared at a pectin concentration of 2% (w/v), pH 7 and calcium content of 10 mmol/l.

Rheological Measurements

Small Deformation Measurements

After preparation, the gelling pectin–calcium mixture was quickly transferred onto the rheometer. Rheological tests were performed by small amplitude oscillatory shear by using a strain controlled rheometer (AR-1500ex, TA Instruments, USA) in oscillatory mode. A plate geometry (5.0 cm in diameter) was used and exposed edges of the sample were covered with mineral oil fluid to prevent evaporation during measurements. Pectin gelation kinetic was monitored at 25°C for 60 min by following the storage (G¢) and loss (G¢¢) modulus. The gel set time (tg, the time at which G¢ and G¢¢ intersected at the study frequency) was also determined. All measurements were carried out at a frequency of 0.25 Hz and 2.5% strain, which was in the linear visco-elastic limit. Frequency sweep (0.01–10 Hz at 2.5% strain) was carried out at the end of the gel formation at 25°C.

Large Deformation Measurements

The texture profile analysis of 6 ml pectin gels made in glass flasks of 30 mm height and 25 mm internal diameter after two hours and 48 hr at 25°C, was registered with Texture Analyzer (TA.XT2, RHEO Stable Micro Systems, Haslemere, England) equipped with XTRAD software version 3.7. The gels were deformed by compression at a constant speed of 1.0 mm/s to a distance of 4 mm from the gel surface using a cylindrical plunger (diameter 25.4 mm) (Carvajal-Millan et al., 2005).

Whippability and Foam Stability

Three pectin solutions were prepared by adding 2.5 g of pectin in 100 ml water at pH values of to 5, 7, and 9 adjusted with NaOH or HCl 4N. Pectin solutions were whipped for 2 min using a hand-held food mixer at a high speed (350 rpm) (Mini MP170 Robot-Coupe) and volumes measured before and after whipping. Whippability was reported as the percentage of increase in volume due to whipping. Foams were slowly transferred to a 1000 ml graduated cylinder and allowed to set at room temperature (21°C) for two hours. Volume of foam after two hours as a percentage of original volume was reported as foam stability (Ramos-Chavira et al., 2009; Temelli, 1997).

Emulsion Stability

Emulsion stabilizing capacity was determined as reported before (Ramos-Chavira et al., 2009; Temelli, 1997). Pectin (0.02 g) was dispersed in 40 ml water at pH 5, 7, and 9. Parafin oil (60 ml) was added slowly and the mixture was blended for 2 min (Brinkmann Polytron homogenizer) at 17,400 rpm and 21°C. Aliquots of emulsion were trasfered into 50 ml graduated tubes and centrifugated for 15 min at 2700 g. Volumes of separated phases were recorded. Emulsion stability was reported as the volume of emulsion remaining unseparated after centrifugation as a percentage of original volume.

Reduced Viscosity

Specific viscosity η_{sp} was determined by measuring the flow time of pectin solutions in an Ubbelohde capillary viscometer (OB size) at 25 ± 0.1°C, immersed in a temperature controlled Koehler bath. Pectin solutions were prepared at different concentrations (0.1–0.5% w/v), dissolving dried pectin in water at pH 5, 7, and 9 with stirring at room temperature. Pectin solutions and solvent were filtered using 0.45 μm membrane filters before viscosity measurements. The η_{sp} was related to the pectin concentration to obtain the reduced viscosity η_{red} (Carvajal-Millan et al., 2005).

STATISTICAL ANALYSIS

All experiments were carried out in triplicate. The experimental data were analyzed with the Statistical Analysis System software (SAS Institute, Cary, NC, USA). The significance of difference was calculated using Tuckey's test ($p \leq 0.05$).

RESULTS AND DISCUSSION

Pectin Extraction and Composition

Yield of pectin extracted from chickpea husk was 8% (w/w) on a dry matter basis (w pectin/w chickpea husk), which is lower than those reported in major sources of pectic substances like apple fruit (16%) (Rascón-Chu et al., 2009). Chickpea husk pectin contains 67% of galacturonic acid, 7.7% of arabinose, 12.3% of galactose, 1.6% of glucose, 0.4% of xylose, 0.6% of mannose, and 10.4% of rhamnose. Chickpea pectin presented an $[\eta]$ of 374 ml/g, which is similar to that reported in LM pectin from yellow passion fruit (Yapo et al., 2006). The viscosimetric molecular weight (M*v*) of chickpea husk pectin was 110 kDa. In HM pectin from low quality apples, Rascón-Chu et al., (2009) found a M*v* value of 112 kDa. From the FTIR spectroscopy analysis, the degree of esterification of chickpea husk pectin was estimated to be 10%.

Gelation

The formation of the pectin gel was rheologically investigated by small amplitude oscillatory shear (Fig. 3.1). The gelation profile followed a characteristic kinetics with an initial increase of G¢ followed by a plateau region. The values of G¢ and G¢¢ at the plateau region (60 min) were 57 and 10 Pa, respectively, which are higher than those reported for other LM pectin gels (Cardoso et al., 2003). It is possible that the attainment of a higher G¢ value in chickpea husk pectin gels in comparison with other LM pectins could be related to longer galacturonic acid blocks within the chain resulting in the formation of higher amounts of "egg-box" structures. The gel set time (tg) was 3.5 min. This behavior is similar to that reported during gelation of LM pectin from olive pomace (Cardoso et al., 2003). In LM pectins, the number of sequences of non-methoxylated galacturonic acid residues is long enough for the formation of the so-called "egg-boxes" resulting in the formation of the gel. Some other intermolecular interactions like hydrogen bonds could be formed, but they are much weaker as compared to the ionic cross-links formed by carboxyl groups (May, 1990). Figure 3.2 shows the mechanical spectrum of pectin gel after 60 min gelation. The mechanical spectrum was typical of solid-like material, with a linear G¢ independent of frequency and G¢¢ much smaller than G¢ and dependent of frequency (Doublier and Cuvelier, 1996). This behavior is similar to that previously reported for commercial LM pectin (DE 23%) (Willats et al., 2006). The tan d (G¢¢/G¢) (data not shown in Fig. 3.2) value calculated at 0.25 Hz was 0.14 for pectin gels, indicating the presence of an elastic system (Ross-Murphy, 1984).

The rheological stability of pectin gels freshly made (two hour) and aged (48 hr) was determined. The gels showed no significant difference on the texture profile analysis (Table 3.1), except for hardness. Twenty-one percent loss in the gel hardness was recorded after storage. This rheological change could be related to an increase in pectin hydration as a result of a moderate hydrolysis of the pectin during storage. Rascón-Chu et al., (2009) reported the texture profile analysis of HM apple pectin gels during storage. At the same pectin concentration, these authors found a lower hardness value (10 g) after gel set but similar hardness loss in the gel (20%) after 48 hr storage at

25°C. Texture evaluation is often an important step in developing a new food product optimizing processing variables.

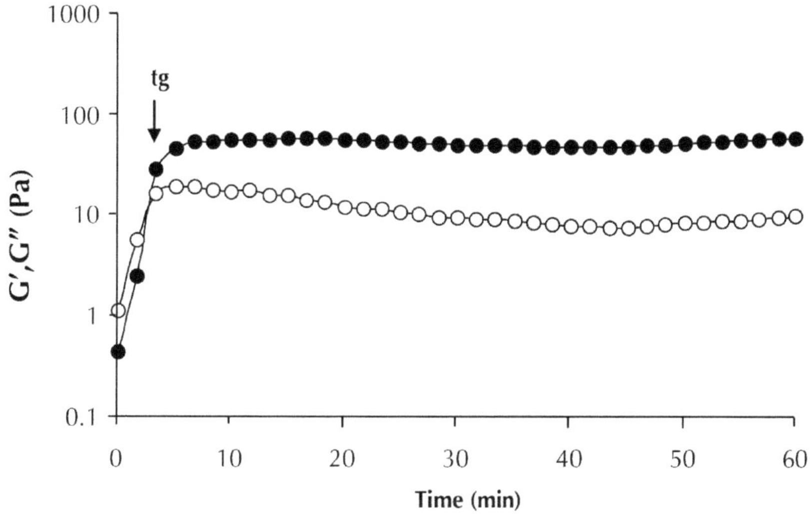

Figure 3.1. Gelation of chickpea husk pectin at 2% (w/v), pH 9 and 10 mmol/l of calcium (G' ●, G" o). Measurements at 25°C, 0.25 Hz and 2.5% strain. tg = gel set time.

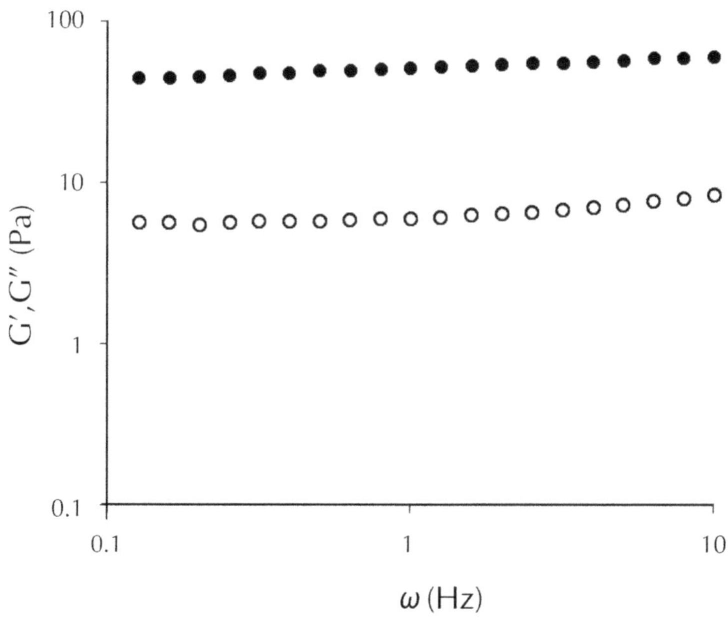

Figure 3.2. Mechanical spectrum (G' ●, G" o) of chickpea husk pectin at 2% (w/v), pH 9 and 10 mmol/l of calcium. Measurements at 25°C and 2.5% strain.

Table 3.1. *Effect of time of storage on texture properties of gels at 2% (w/v) in pectin, pH=9 and 10 mmol/l of calcium.

	After gel set (two hours)	*After gel storage (48 hr)*
Hardness (g)	99[a]	78[b]
Adhesiveness	–295[a]	–278[a]
Springiness	0.93[a]	0.95[a]
Cohesiveness	0.79[a]	0.84[a]
Gumminess	78[a]	76[a]
Resilience	0.04[a]	0.06[a]

*Means in a row with different letters are significantly different ($p \leq 0.05$).
All results are obtained from triplicates.

Functional Properties

Whippability, foam and emulsion stability, and reduced viscosity of pectin dispersions at different pH values are presented in Table 3.2. Whippability significantly increased from 104–124% as the pH increased from 5 to 9. Foam and emulsion stability significantly increased from 87–95% and from 75–96%, respectively, when the pH augmented from 5–9. Nevertheless, no significant difference was found in foam and emulsion stability at pH 7 and 9. This increase in whippability and foam and emulsion stability could be related to the fact that pectin solution reduced viscosity (η_{red}) values increased from 57 to 61 as the pH changed from 5 to 9. Polysaccharides contributed to the stability of foam and emulsion systems, mainly by increasing the viscosity of the aqueous phase. They do not interact with the hydrophobic phase, since they are not true surfactants. Coalescence of air bubbles and oil droplets in foam and emulsion systems, respectively, is hindered by a viscous aqueous phase (Temelli, 1997).

Table 3.2. *Functional properties of chickpea husk pectin.

pH	Whippability (%)	Foam Stability (%)	Emulsion stalbility (%)	Reduced viscosity (ml/g)
5	104[c]	87[b]	75[b]	57[c]
7	112[b]	93[a]	96[a]	59[b]
9	124[a]	95[a]	96[a]	61[a]

*Means in a column with different letters are significantly different ($p \leq 0.05$).
All results are obtained from triplicates.

CONCLUSION

Pectin can be recovered from chickpea husk. This is low methoxy pectin capable of forming elastic gels by calcium addition. Chickpea husk pectin gels present a higher G¢ value in comparison to other LM pectins, which could be related to longer galacturonic acid blocks within the chain resulting in the formation of higher amounts of "egg-box" structures. After storage, pectin gels showed no significant difference on the texture profile analysis, except for a 20% hardness loss. Whippability, foam and emulsion stability, and viscosity increase as the pH changes from 5 to 9. The results suggest that chickpea husk could be a potential source of gelling pectin for food applications. Further research is undergoing in order to explore the structural properties of this hydrocolloid.

FUTURE CONSIDERATIONS

Chickpea husk as a source of low methoxy pectin continues to be investigated and new information about their potential application in the food industry is being generated. Nevertheless, more research is needed to elucidate several questions, especially those concerning the structure of this polysaccharide. Additional studies will also be required in order to understand the effect of temperature, pH, pectin, and calcium concentration on the gel properties.

ACKNOWLEDGMENTS

This research was supported by Sagarpa/Conacyt/Cofupro (grant 48735 to A. Rascon-Chu, PhD). The authors are pleased to acknowledge Francisco Vazquez-Lara and Maria del Carmen Granados-Nevárez for their technical assistance.

KEYWORDS

- **Functional**
- **Gelation**
- **Low methoxy pectin**
- **Rheology**
- **Texture**
- **Viscosity**

Chapter 4

Total Phenolic, Flavonoids, Tannin Content and Antioxidant Activity of Dried Plants *Garcinia mangostana* Linn. and *Garcinia atroviridis Griff.* ex T. Anders

Hasnah, O., Afidah, A. R., Rizal, R., Nornaemah, M. B., Nurulafiqah, M. T., and Nor Hafizah, N.

INTRODUCTION

Phenolic, flavonoid, and tannin content of dried *G. mangostana* and *G. atroviridis* were investigated using the well established method, via Ultraviolet-visible (UV-Vis) Spectrometry. Most of the sample extracts of both plants have higher phenolic content, followed by flavonoid and tannin content. Antioxidant activities of dried samples of *G. mangostana* and *G. atroviridis* were carried out by using 2, 2-diphenyl-1-picrylhydrazyl (DPPH) assay. *G. mangostana* sample extracts have higher antioxidant activity than *G. atroviridis*. *G. mangostana*'s leaves in methanol extract showed the highest inhibition percentage (95.72%) followed by *G. mangostana*'s pericarp (95.43%). *G. atroviridis*'s stems in the hexane extract showed the lowest inhibition percentage (51.88%). No correlation was observed between the total phenolic, flavonoid, tannin content, and the antioxidant activity.

Oxidation is basic part of aerobic life and our metabolism. Thus, radicals that are produced from this reaction contribute harmful effects such as peroxidation of membrane lipids, aggression to tissue proteins and membranes, damage to DNA and enzymes (Husain et al., 1987). Therefore, studies of antioxidants in foods and cosmetics are becoming popular these days mainly due to the findings of the free radical's effect in organisms. Antioxidants play an important role in delaying the oxidation of other molecules by inhibiting the initiation or propagation of oxidizing chain reactions by free radicals (Namiki, 1990). The risk of diseases can be reduced by increased consumptions of antioxidants that are abundant in food. Although well known antioxidants such as vitamin C, vitamin E, and β-carotene have been attributed well, the naturally occurring compounds may also be significant as antioxidant.

Generally, *Garcinia* species is from the family of *Clusiaceae. Garcinia mangostana* L. (Mangosteen) is a tropical evergreen tree that may be found in Malaysia, India, Thailand, Vietnam, Singapore, Philippines, and Burma. *G. mangostana* is a natural remedy for abdominal pain, dysentery, diarrhea, suppuration, infected wound, leucorrhoea, chronic ulcer, and gonorrhea (Jayaprakasha et al., 2006). *G. mangostana* exhibits an anti-inflammatory (Gopalakrishnan et al., 1997), antitumor and antioxidant abilities (Williams et al., 1995), and it also has potential as antibacterial activity against *Staphylococcus aureus* (Sakagami et al., 2005) and *Helicobacter pyroli* (Mahabusarakum et al., 1983). It was reported that the hull of *G. mangostana* was a source of mangostin, tannin, xanthone, isoflavone, flavone, and other bioactive substances (Deachathai et al., 2005; Deachathai et al., 2006).

Garcinia atroviridis, also known as Asam Gelugor, is common in the Malay Peninsula. Its fruits are cultivated throughout Southeast Asia, for spices. *G. atroviridis* also has been reported to be effective in weight management on a short-term basis as the fruits contain fruiting acids such as hydroxycitric acid (HCA) that promotes fat burning and spares carbohydrate utilization at rest or during exercise (Sullivan et al., 1974), help to promote weight loss by lowering lipogenesis and increasing glycogen development, thus decreasing appetite (Kovacs et al., 2001). Diet rich in antioxidants provides protection against lipid peroxidation and free radical generation, and also inhibits the development of atherosclerosis (Singh et al., 1995). Hence, *Garcinia atroviridis* may be useful in preventing atherosclerosis or lowering the relative risk of atherosclerosis. The crude extracts of fruits and leaves of *G. atroviridis* exhibited significant antifungal activity against *Cladosporium herbrum* (Mackeen et al., 2000). *Garcinia atroviridis* also have afforded the isolation of γ-lactone, atroviridin, atrovirisidone, and atrovirinone (Kosin et al., 1998; Lewis and Neelakantan, 1965; Permana et al., 2001) as well as the identification of some organic acids, pentadecanoic, octadecanoic, nonadecanoic, and dodecanoic acids in its fruit by GC-MS).

Plant polyphenols are large groups of secondary metabolites ranging from simple molecules such as phenolic acids to highly polymerized constituents such as tannins. Naturally occurring compounds such as phenolic, flavonoid, and tannin compounds are reported to exhibit antioxidant properties (Bors and Saran, 1987; Emmons and Peterson, 2001; Konczak and Zhang, 2004), and act as remedy to some serious diseases. The antioxidant activity of phenolic compounds might be due to their redox properties that lead them to be excellent reducing agents, hydrogen donor, singlet oxygen quencher, and metal chelator (Rafael et al., 2008). Phenolic compounds also play important roles in plant resistance and defense against microbial infections, which is due to the reactive oxygen species (ROS) (Grassman et al., 2002). Flavonoids play important pharmalogical roles against various human diseases, such as cardiovascular disease, cancer, inflammation, and allergies (Jiangrong and Yueming, 2007). Tannin is usually related to the pigmentation that exists as plant condensed tannins (anthocyanidin) which have radical-scavenging activity, antioxidant activity, and anti-tumor activity (Nour-Eddine et al., 2007). Anthocyanidin is a basic structure of the anthocyanin, which is reported to have a vital role in the prevention of neuronal and cardiovascular illness cancer, diabetes (Konczak and Zhang, 2004), and in human nutrition (Stintzing and Carle, 2004) and its biological activity (Kong et al., 2003). Numerous positive health effects have been described, such as biological activities including radical scavenging, anti-inflammatory, anti-mutagenic, anti-cancer, anti-HIV, anti-allergic, anti platelet, and antioxidant activity (Block, 1992; Gabor, 1986; Middleton et al., 2000). Some journals reported the correlation of these biological activities with the plant's chemical content, such as phenolics, flavonoids, and tannins. The aim of this study to determine the total phenolic, flavonoid, and tannin content of dried *G. mangostana* and *G. atroviridis* plants, and to evaluate the antioxidant activities of both plants.

MATERIALS AND METHODS

Chemicals and Reagents

Quercetin and 2, 2-diphenil-1-pycrihydrazyl (DPPH) were purchased from Sigma Chemical Co., USA. Folin-Ciocalteu 2 N, butylated hydroxytoluene (BHT), and (+)-Catechin hydrate were purchased from Sigma-Aldrich and methanol, hexane, acetone, and ethyl acetate were purchased from Merck Analysis Grade. All the solvents and chemicals used in this study were of analytical grade or HPLC grade.

Plant Materials

Garcinia mangostana and *Garcinia atroviridis* were obtained from Changlun, Kedah, Malaysia. The plant's parts used in this investigation for *Garcinia mangostana* were leaves, stems, and pericarps whereas leaves and stems were used for *Garcinia atroviridis*.

Preparation of Extracts

The dried samples were air dried at room temperature (30˚C) for 14 days until a constant weight was achieved. Plant materials (100g) of the crude dried plants were extracted three times with methanol. Then, the methanol crude was extracted with hexane followed by ethyl acetate. Methanol, hexane, and ethyl acetate extracts were used for subsequent experiments.

Measuremant of Total Phenolic Content (TPC)

Total phenolic content in both plants estimated by using Folin–Ciocalteu assay according to Vidushi et al., (2006). A volume of 0.25 ml (100 ppm) sample, 3.75 ml of Folin–Ciocalteu 2 N reagent (1:15; Folin-Ciocalteu 2 N reagent:distilled water) were put in a vial. After 3 min 1.25 ml of Na_2CO_3 was added and shaken thoroughly and was left in the water bath (40˚C) for 30 min. Then, TPC of the samples was measured using UV-Vis spectrophotometer at 685 nm. The results were expressed as mg of Catechin per gram of sample, and mg of Quercetin per gram of sample.

Measurement of Total Flavonoid Content (TFC)

The total flavonoid content of each extract was measured according to Aline et al. (2005). A volume of 5 ml of sample and 5 ml of 2% $AlCl_3$ were placed in vials and incubated for 5 min. at room temperature (30°C). Then, TFC of the samples were measured using spectrophotometer at 415 nm. The results were expressed as mg of Catechin per gram of sample, and mg of Quercetin per gram of sample.

Measurement of Total Tannin Content (TTC)

The total tannin content of each extract was measured according to Sze-Tao et al. (2001). A volume of 3 ml sample, 3 ml of 4% vanillin in methanol, 1.5 ml 37% HCl were put in vials and were incubated in the dark for 10 min. Then, TTC of the samples were measured using UV-Vis spectrophotometer at 500 nm. The results were expressed as mg of Catechin per gram of sample, and mg of Quercetin per gram of sample.

DPPH Free Radical Scavenging Assay

Radical scavenging activities of the extracts were assayed according to a method described by Afidah et al. (2008). A volume of 4 ml sample extract of various concentrations, and 0.5 ml DPPH were placed in vials covered with aluminium foil, and incubated for 30 min at room temperature (30°C) before being measured at 517 nm. Quercetin, catechin, α-tocopherol, and BHT were used as references. The percentage of inhibition sample extracts was measured by the following formula:

$$\%\ \text{Inhibition} = \frac{(A_{\text{blank}} - A_{\text{sample}}) \times 100\%}{A_{\text{blank}}}$$

where A_{blank} is absorbance of reaction, without sample extracts, and A_{sample} is absorbance of reaction with sample extracts.

Measurement of IC_{50}

The IC_{50} is a measure of the effectiveness of a compound in inhibiting biological or biochemical function or in other words, it is the half maximal (50%) inhibitory concentration (IC) of a substance (50% IC, or IC_{50}).

RESULTS AND DISCUSSION

The percentage yield of all extracts were between 0.19–16.01% w/v. The total phenolic, flavonoid, and tannin content and antioxidant activities varied among sample extracts. The total phenolic, flavonoid, and tannin content are shown in Table 4.1. The results demonstrated that all sample extracts have higher phenolic compounds, than flavonoids and tannins. Among all sample extracts, ethyl acetate extracts showed the highest phenolic, flavonoid, and tannin content compared to the methanol and hexane extracts (see Table 4.2).

Table 4.1. Total phenolic content (TPC), total flavonoid content (TFC), total tannin content (TTC).

	TPC (mg/g)		TFC (mg/g)		TTC (mg/g)	
	Catechin	Quercetin	Catechin	Quercetin	Catechin	Quercetin
GA-L-EA	113.16 ± 1.99	100.86 ± 7.95	23.48 ± 0.48	23.41 ± 0.80	5.91± 0.08	0.06 ± 0.02
GA-L-H	103.33±4.89	92.38±2.12	47.28±2.92	47.21±0.45	25.00±1.24	13.19±0.84
GA-L-ME	102.63±2.99	91.77±2.82	1.66±0.60	1.59±0.18	6.36±0.22	0.38±0.01
GM-L-EA	51.53±0.33	55.67±0.38	21.52±1.33	21.59±1.04	4.44±0.80	11.82±1.82
GM-S-EA	49.91±1.80	53.81±2.61	27.38±2.57	27.45±0.63	5.69±1.27	13.64±1.03
GM-L-ME	35.37±0.01	37.07±0.13	19.45±1.41	19.52±0.73	3.50±0.04	10.45±0.22
GM-P-EA	35.13±3.03	36.79±1.75	7.38±0.25	7.45±0.35	1.00±0.08	6.82±1.54
GM-S-ME	21.97±1.80	21.81±2.61	17.21±2.57	17.28±0.63	3.19±1.27	10.00±1.03
GA-S-EA	20.00±3.03	20.41±2.77	0.34±0.12	0.28±0.02	7.73±0.45	1.31±0.29

Table 4.1. *(Continued)*

	TPC (mg/g)		TFC (mg/g)		TTC (mg/g)	
	Catechin	**Quercetin**	**Catechin**	**Quercetin**	**Catechin**	**Quercetin**
GM-P-ME	19.73±1.72	19.07±1.87	11.00±0.54	11.07±0.20	2.25±0.99	8.64±1.52
GM-P-H	19.36±2.65	18.65±2.87	6. 00±0.37	6.07±0.02	0.38±0.11	5.91±0.95
GM-S-H	16.31±0.22	15.14±0.98	7.03±0.26	7.10±0.81	0.38±0.10	5.91±0.19
GA-S-ME	15.26±2.28	16.32±2.90	0.90±0.34	0.83±0.01	7.73±2.38	1.31±0.08
GM-L-H	14.95±0.22	13.68±0.98	10.48±0.26	10.55±0.81	1.94±0.10	8.18±0.19
GA-S-H	9.12±0.34	11.02±0.63	0.28±0.01	0.21±0.01	9.09±2.23	2.25±0.14

Table 4.2. Percentage of inhibition.

Sample extracts	Inhibition (%)
QUERCETIN	96.79
GM-L-ME	95.72
GM-P-ME	95.43
GM-P-EA	95.27
GM-L-EA	95.05
GM-S-ME	94.95
GM-S-EA	93.07
GM-P-H	91.65
GM-S-H	90.63
CATECHIN	84.79
α-TOCOPHEROL	79.92
GM-L-H	70.86
BHT	61.53
GA-L-EA	59.18
GA-L-ME	57.97
GA-L-H	55.67
GA-S-EA	54.96
GA-S-ME	54.75
GA-S-H	51.88

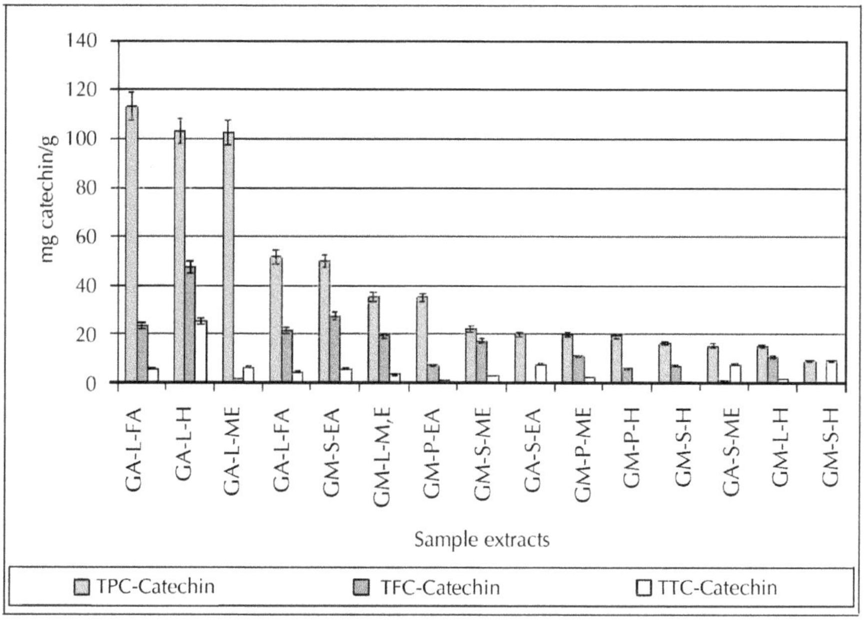

Figure 4.1. Graph comparison between TPC, TFC and TTC (Catechin).

Abbreviations: GM.L., *G. mangostana's* leaves; GM.S., *G. mangostana's* s stem; GM.P., *G. mangostana's* s pericarp; GA.L., *G. atroviridis's* leaves; GA.S., *G. atroviridis's* stem; H, Hexane; EA, Ethyl acetate; ME, Methanol

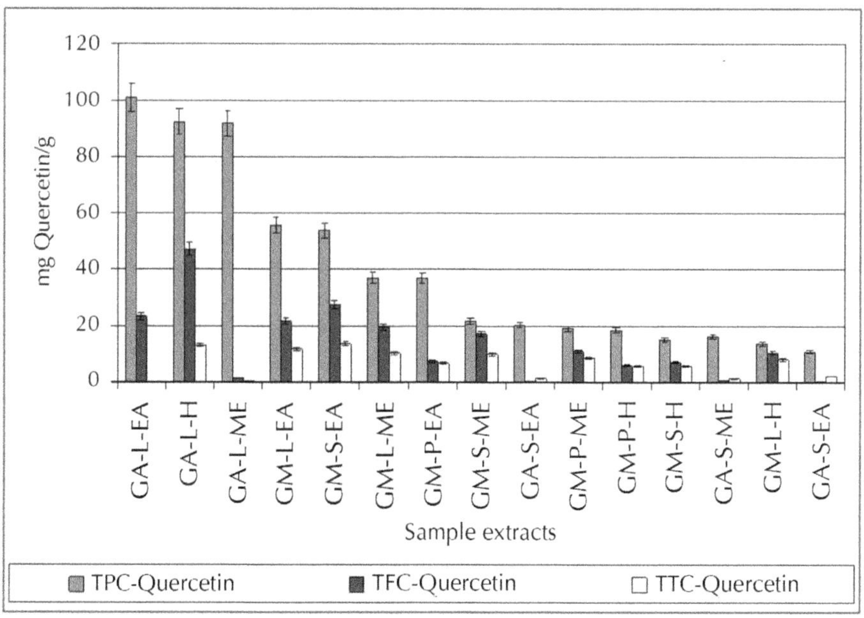

Figure 4.2. Graph comparison between TPC, TFC and TTC (Quercetin).

The antioxidant activity indicated the ability of the sample extracts to neutralize free radical compounds. The deep violet color of DPPH reduced to pale yellow color showing that the free radical is accepting hydrogen atom from the hydrogen donor.

Figure 4.3 shows the percentage of inhibition for all sample extracts, as well as the references (quercetin, catechin, α-tocopherol, and BHT). The sequence of the percentage of inhibition for all sample extracts, including the references at 100 ppm, is as follows:

Quercetin > GM-L (ME) > GM-P (ME) > GM-P (EA) > GM-L (EA) > GM-S (ME) > GM-S (EA) > GM-P (H) > GM-S (H) > Catechin > α-tocopherol > GM-L (H) > BHT > GA-L (EA) > GA-L (ME) > GA-L (H) > GA-S (EA) > GA-S (ME) > GA-S (H).

The trends are in accordance with the polarity of solvents.

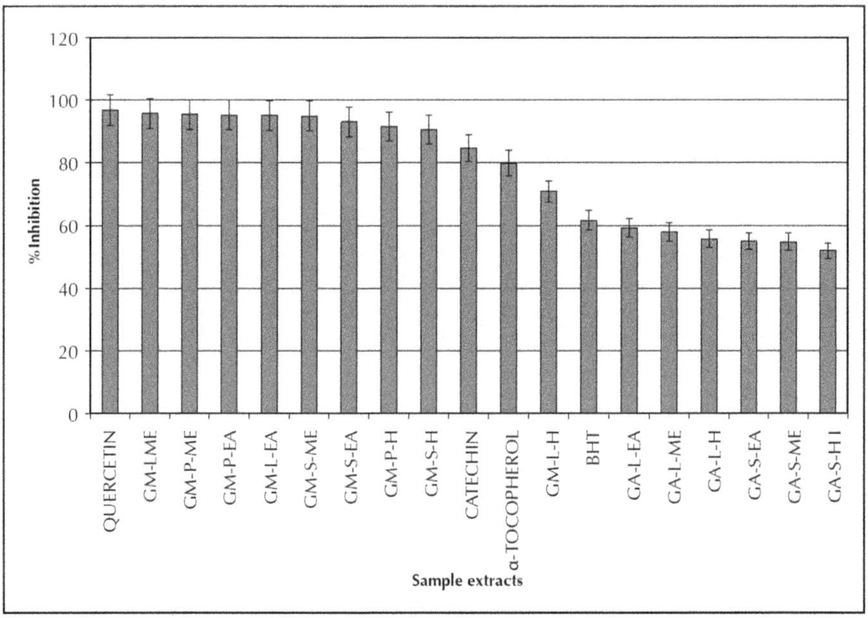

Figure 4.3. Graph DPPH's scavenging acivity.

Quercetin is an example of a flavonoid compound. Since flavonoid had been reported to exhibit great potential antioxidant activity (Orsolya et al., 2004), the phenomenon (Quercetin has highest inhibition's percentage among others) in this study is acceptable. From this study, all *G. mangostana*'s sample extracts have better antioxidant activity, than the reference, BHT. *G. mangostana*'s leaves in the methanol extract have the highest percentage of inhibition, indicating its potential as an antioxidant. On the other hand, *G. atroviridis*'s stems in the hexane extract have the lowest percentage of inhibition, consistent with the work of Kalyarat and Kaew (2006) and Nor Hadiani and Khozirah (2008).

The sample is said to have a high, moderate, and low antioxidant activity, if the inhibition percentage is > 90%, 60–90% and < 60%, respectively (Kalyarat and Kaew,

2006). From this study, all *G. mangostana*'s sample extracts have high antioxidant activity, except the leaves from the hexane extract which has a moderate potential antioxidant activity. The *G. atroviridis*'s sample extracts however exhibited a low antioxidant activity, as the percentage of inhibition was below 60%. This could be due to the instability of the phenolic, flavonoid, and tannin compounds (Amin et al., 2004; Giusti and Wrolstad, 2001; Zadernowski et al., 2005). In addition, the temperature change during the drying process might cause the degradation of some of the polyphenol substituents (Sri et al., 2008). Climatic growth conditions, growth, duration of storage, and the cultivation of the plant may also influence the antioxidants activity (Papetti et al., 1998).

The sequence of the sample extracts, and the reference for their IC_{50} are as follows:

GA-S (H) > GA-S (ME) > GM-L (H) > GA-L (H) > GA-S (EA) > GM-P (H) > GM-S (H) > GA-L (EA) > BHT > α-tocopherol > GA-L (ME) > Catechin > GM-S (EA) > GM-P (EA) > Quercetin > GM-L (EA) > GM-S (ME) > GM-P (ME) > GM-L (ME).

A lower value of IC_{50} indicates a higher effectiveness in inhibiting biological or biochemical functions, leading to a higher antioxidant activity (Nor Hadiani and Khozirah, 2008). According to the above sequence, *G. mangostana*'s leaves in methanol extract showed the highest antioxidant activity, while the *G. atroviridis*'s stems in the hexane extract showed the lowest antioxidant activity.

A comparison of the total phenolic, flavonoid, and tannin content with the antioxidant activity is shown in Figs. 4.4–4.6, respectively. There are no correlations between the antioxidant activity and the total phenolic, flavonoid, and tannin content, as similarly observed by Kähkönen et al., (1999) on the correlation between phenolic content and antioxidant activity.

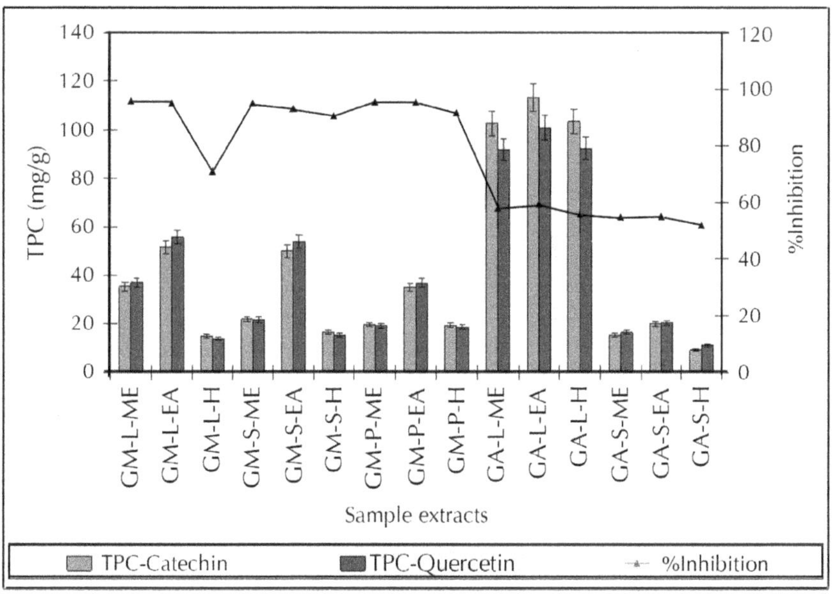

Figure 4.4. Graph comparison between % Inhibition and TPC (Catechin and Quercetin).

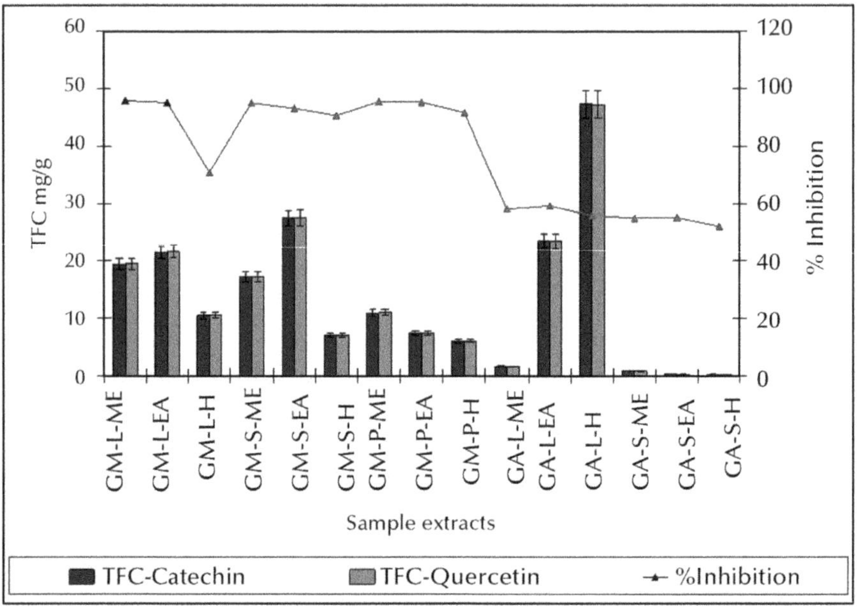

Figure 4.5. Graph comparison between % Inhibition and TFC (Catechin and Quercetin).

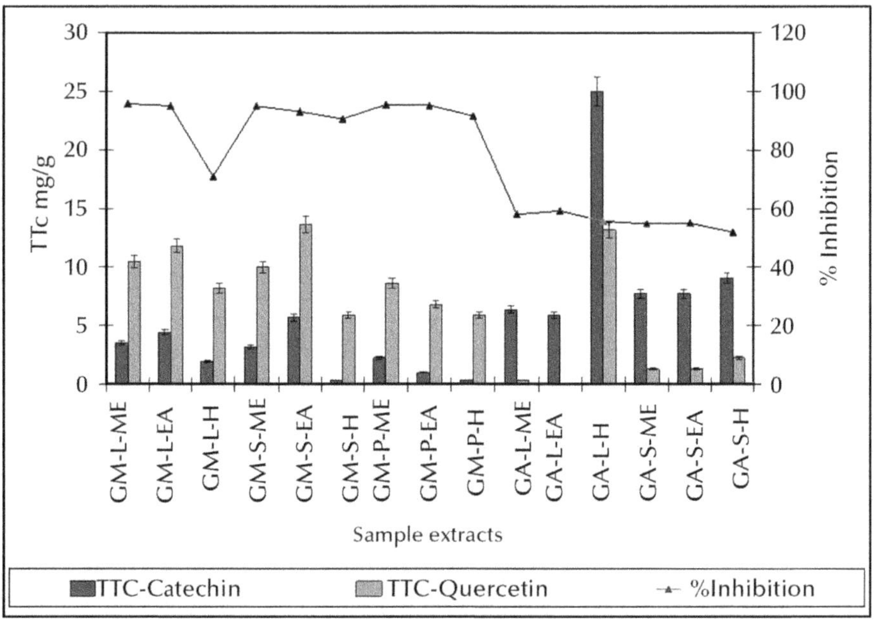

Figure 4.6. Graph comparison between % Inhibition and TTC (Catechin and Quercetin).

CONCLUSION

Ethyl acetate extract for all samples gave the highest total phenolic, flavonoid, and tannin content compared to the methanol and hexane extracts. Most of the sample extracts of dried *G. mangostana* showed high antioxidant activities while that of *G. atroviridis* showed low antioxidant activities. This study also showed that the leaves of both plants had higher antioxidant activity than the pericarp and stems. No correlation was observed between the total phenolic, flavonoid, and tannin content and the antioxidant activity.

ACKNOWLEDGEMENTS

The authors would like to express their appreciation to Universiti Sains Malaysia, for the financial support given through the Grant Research University (1001/PKIM-IA/811133).

KEYWORDS

- **2, 2- Diphenyl-1-picrylhydrazyl (DPPH)**
- **Flavonoid**
- ***Garcinia atroviridis***
- ***Garcinia mangostana***
- **Phenolic**
- **Tannin**

Chapter 5

Procyanidins Improve Some Disrupted Glucose Homoeostatic Situations: An Analysis of Doses and Treatments According to Different Animal Models

Montserrat Pinent, Mayte Blay, and Anna Ardévol

INTRODUCTION

This review analyzes the potential beneficial effects of procyanidins, the main class of flavonoids, in situations in which glucose homeostasis is disrupted. Because the disruption of glucose homeostasis can occur as the result of various different causes, we critically review the effects of procyanidins based on the specific origin of each type of disruption.

Where little or no insulin is present (type I diabetic animals), summarized studies of procyanidin treatment suggest that procyanidins have a short-lived insulin-mimetic effect on the internal targets of the organism—an effect not reproduced in normoglycemic, normoinsulinemic healthy animals. A very different situation is insulin resistance (usually linked to hyperinsulinemia). Preventive studies using fructose-fed models indicate that procyanidins may be useful in preventing the induction of damage and thus in limiting hyperglycemia. But the results of other studies using models such as high-fat diet treated rats or genetically obese animals are controversial. Although the effects on glucose parameters are hazy, it is known that procyanidins target key tissues involved in its homeostasis. Interestingly, all available data suggests that procyanidins are more effective when administered in one acute load than when mixed with food.

Abnormalities affecting glucose homeostasis, that is, in insulin secretion and/or insulin action, lead to the progressive deterioration of glucose tolerance and cause hyperglycemia, which in turn leads to a complex metabolic disorder of the endocrine system called diabetes mellitus (type II diabetes). The disease is a major public health problem found in all parts of the world and its incidence is rapidly increasing (Bajaj and DeFronzo, 2003), which has led to a growing interest in the search for hypoglycemic agents from natural products, especially those derived from plants.

Flavonoids are naturally occurring phenolic compounds that are commonly found in plants. They are in fruits and vegetables, in chocolate, as well as in drinks such as wine and tea (Aron and Kennedy, 2008; Bhagwat et al., 2010). Their basic structure consists of three phenolic rings (Aherne and O'Brien, 2002). Depending on the structure and oxidation level of the C-ring, flavonoids are further divided into several subclasses (Fig. 5.1). So, flavonoids include thousands of structures with different chemical, physical, and biological properties which perform a wide array of biochemical and pharmacological actions such as cardioprotective (Zern and Fernandez, 2005), antioxidant (Williams and Spencer, 2004), anti-inflammatory (Rivera et al., 2008), antiallergic, antiviral and anticarcenogenic (Amin and Buratovich, 2007; Kim et al., 2008) activities. Among their effects, several monomeric flavonoids and natural extracts rich in monomeric forms have been shown to improve hyperglycemia in streptozotocin-induced diabetic rats, in genetically altered diabetic mice, and in animal models with diet-induced insulin resistance or diabetes (Bose et al., 2008; Cao et al., 2007; Diepvens et al., 2006; Park, 2008; Potenza et al., 2007).

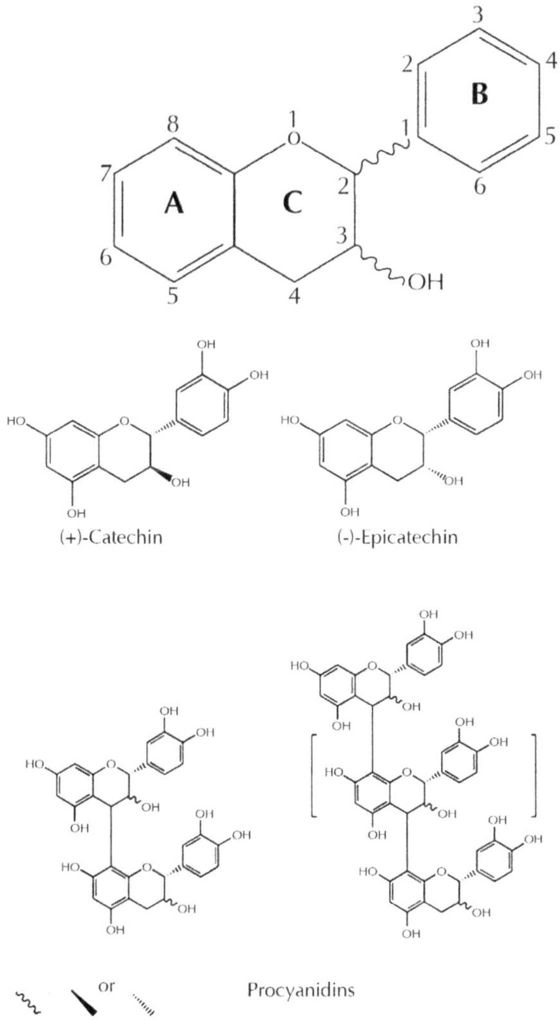

(a)

(b)

Figure 5.1. Representative structure of procyanidins: (a) Basic structure of a flavonoid and (b) Structure of the flavanol monomers (+)-catechin and (–)-epicatechin, and their polymers called procyanidins, which include from dimers to oligomers.

This review focuses on the effects of the largest and most ubiquitous class of flavonoids, the flavan-3-ols, which comprise the main constitutive units of condensed procyanidins. Homo-oligomeric proanthocyanidins with two (3′,4′) B-ring hydroxyl groups are called procyanidins and are one of the most common types of proanthocyanidins found in nature (Aron and Kennedy, 2008). Procyanidins are oligomeric structures formed by the polymerization of the monomeric flavan-3-ols (+)-catechin and (–)-epicatechin up to ten subunits. Procyanidins are found in a wide variety of foods in the human diet. They are present in the fruits, bark, leaves, and seeds of many plants and plant-derived foods such as green tea, apples, cocoa, chocolate, grapes, apricots, and cherries. They are especially abundant in fruit juices and red wine (Aherne and O'Brien, 2002; Aron and Kennedy, 2008). Several authors have studied flavan-3-ol consumption and bioavailability, although the lack of reliable concentration data for procyanidin in foods has made it difficult to accurately evaluate their dietary intake. Gu et al. reported that daily proanthocyanidin intake could vary from 10 mg to 500 mg/day and they estimated that the mean daily proanthocyanidin intake in the United States was 53.6 mg/person/day (Gu et al., 2004). Arts et al. estimated that the mean intake of flavan-3-ol monomers in the Netherlands was 50 ± 56 mg/day, although the intake of oligomeric forms was not analyzed in that study (Arts and Hollman, 2005). Due to the considerable intake of procyanidins through the diet, their potential beneficial effects have been widely studied. Procyanidins act against coronary heart diseases and atherosclerosis, as well as several metabolic processes associated with the development of those disorders (Bladé et al., 2010). They are involved in the modulation of cholesterol and lipid metabolism (Bladé et al., 2010), induce changes in vascular events (Corder et al., 2006), have antigenotoxic (Llópiz et al., 2004) and cardiovascular effects (Karthikeyan et al., 2007) and improve the oxidative or inflammatory states (Roig et al., 1999; Roig et al., 2002; Puiggros et al., 2009; Terra et al., 2009). They also have antiproliferative effects and have been studied as cancer preventive agents (Faria et al., 2006; Mantena et al., 2006). Several studies have investigated their effects on glucose homeostasis-disrupted situations; however a clear consensus has not been reached on what exactly those effects are. Some evidence suggests that they might act as hypoglycemic agents, but hyperglycemia can result from various different causes. On the one hand, it might be the consequence of type I diabetes, in which there is a loss of β-cells due to an immune assault (Ichinose et al., 2007). On the other hand, high glucose levels can occur due to the ineffectiveness of insulin, that is insulin resistance, a condition in which insulin levels initially increase to compensate for the lack of effect before reaching a state in which the pancreas is no longer functional, which is type II diabetes (Kashyap and DeFronzo, 2007). Considering, therefore, that the causes that lead to glucose homeostasis disruption have different molecular origins, one could speculate that the effects of procyanidins might depend on the specific condition of glucose homeostasis disruption in which they are analyzed. Consequently, it is important to take into account all these situations before making general statements about these compounds. In this review we summarize the effects of procyanidins and their possible mechanisms of action, critically analyzing their role in different glucose homeostasis situations.

Procyanidin Bioactivity in Insulin Deficiency

Hyperglycemia in type I diabetes is the result of the body's inability to synthesize and/or secrete functional insulin. Several authors have assayed the ability of procyanidin-enriched extracts to ameliorate the physiological state caused by this situation (summarized in Table 5.1), which is easily reproduced with animal models through the destruction of the pancreas, mainly by means of streptozotocin (STZ).

In this model, the lowest dose of procyanidins that has been assayed, as far as we know, is 10 mg/kg body weight (bw). Maritim et al. (2003) showed that at this dose a procyanidin-rich extract from pine bark (Pycnogenol®) improved serum glucose after 14 days of treatment. This is the only model in which procyanidins were administered intraperitoneally; in other studies they are given orally, thus more closely reproducing the physiological ingestion of a procyanidin extract, since it can be digested, metabolized, not fully absorbed, and so forth (Lee et al., 2007) also administered 10 mg/kg bw of procyanidins orally once a day for three weeks and they describe an improvement in serum glucose and glycosylated protein due to oligomeric forms. In the same study the authors showed that polymeric procyanidins do not have the same effect. A similar dose of extract (in this case, polyphenol-enriched white wine and its derivative free ethanol) administered twice daily for a period of six weeks was used by (Landrault et al., 2003), who found a tendency toward a decrease in plasma glucose, although not statistically significant. In that case, the dose of procyanidins was not given all at once in a daily bolus, but was added to feed pellets used in routine animal feeding. Similarly, in a study by (Osakabe et al., 2004) a dose of roughly 50 mg/kg bw (in fact, it is difficult to clearly identify the administered dose in that study) of polyphenols in the form of cocoa liquor for a period of 10 weeks did not show any significant effect either on plasma glucose or on body weight. However, slight improvements were found for the same parameters in the procyanidins-treated group. Al-Awwadi et al. tested a procyanidins dose of roughly 200 mg/kg bw (Al-Awwadi et al., 2004), which showed a clearer antihyperglycemic effect. This effect of reducing blood glucose was only found during daily measurements, whereas when measured during fasting, without a previous daily administration of procyanidins no effects on glycemia were recorded. Considering this, along with the fact that insulin levels were not modified in any of the procyanidin treatments, the authors explain that the effects of procyanidins are related to the reduction of food intake and/or absorption of nutriments and suggest a short-lived effect not involving the correction of the diabetic state. Our group worked with a similar amount (250 mg/kg) of a grape seed-derived extract enriched with oligomeric procyanidins (GSPE) administered acutely (five hours of treatment), which limited hyperglycemia in male STZ-diabetic Wistar rats in a fed state (Pinent et al., 2004). A parallel study with fasted animals showed that the reduction of glucose absorption was not the main mechanism used by procyanidins to decrease glycemia (Pinent et al., 2004). In a further analysis, we looked at the effects of the same dosage of GSPE in an oral sucrose tolerance test (supplementary data: Fig. 5.1) where we found no delay in the appearance of glucose in the blood in GSPE treated animals compared to controls, suggesting that grape procyanidins do not act primarily (or only) through disturbing glucose absorption. Instead, our results suggest that the antihyperglycemic

Table 5.1. Summary of published effects of oligomeric procyanidins on glucose homeostasis in type I diabetic animal models.

Estimated Dose*	Published dose	Extract	Duration	Animal model	Glucose measurement	Effects on glucose	References
	10 mg/kg BW; daily; intraperitoneally	Pycnogenol, a standarized extract from French maritime pine bark	14 days	Female Sprague-Dawley STZ-diabetic rat	On day 15, cardiac puncture halothane anaesthesia	Decreased serum glucose	Maritim et al., 2003
	10 mg/kg BW; daily; forced orally	Oligomeric procyanidins (MPD: 3.3)/polymeric from persimmon peel	3 weeks	Male Wistar STZ-diabetic rat	24 hr after last dose (not clear) pentobarbital anaesthesia	Decreased serum glucose and glicosilated protein	Lee et al., 2007
14.25 mg "gallic eq"/kg twice daily	10 mg/kg BW; twice daily	Polyphenol enriched white wine (no anthocyanins); with and without etanol	6 weeks	Male Wistar STZ-diabetic rat	Not clear	Tendency to decrease glycaemia. Improved oxidative stress status	Landrault et al., 2003
Around 50 mg polyphenol/kg bw in the pellet	Diet containing 0.5% CLP (2.2% catechin+ epicatechin; 1.69% B2; 2.37% C1; 2.01% cinnamtannin A2)	Polyphenol derived from cacao liquor	10 weeks	Male Sprague-Dawley STZ-diabetic rat	Some hours later than the polyphenols intake;from tail vein	No effects on glucose	Osakabe et al., 2004
	200 mg "gallic eq"/kg bw in a single dose at 9 a.m.	Red wine extract	6 weeks	Male Wistar STZ-diabetic rat	Monthly simultaneous to the procyanidins dose. End experiment and OGTT, 24 hr after last dose	Decreased glucose linked to the extract administration	Al-Awwadi et al. 2004
	250 mg/kg bw forced orally	GSPE: monomeric (21.3%), dimeric (17.4%), trimeric (16.3%), tetrameric (13.3%), and oligomeric (5–13 units) (31.7%)	1 dose	Male Wistar control and STZ-diabetic rat	5 hr after the dose	decrease glucose	Pinent et al., 2004
	250 mg/kg bw forced orally	GSPE	24 weeks	Male Wistar STZ-diabetic rat	After overnight fasting	No changes in glucose or glucosilated hemoglobin	Li et al., 2008
	50 and 100mg/kg bw; once daily; oral gavage	GSPE	At 24, 48 and 72 hr	Male Wistar alloxan-diabetic rat	After 18 hr fasting	Decreased glucose (dose 100 after 48 hr and both doses after 72 hr) decreased insulin at 72 hr	El-Alfy et al., 2005

*In order to compare the different studies we show the administered doses as mg procyanidins/kg body weight, therefore for the studies in which other units were used, we have estimated the amount of mg procyanidins/kg body weight from the published data.

effect of procyanidins is in part due to the fact that procyanidins have central insulin-mimetic effects. They favor glucose uptake in the liver through increasing the mRNA expression of glucokinase (Gck) and decreasing the mRNA of glucokinase-regulatory protein (Gckr) and glucose-6-phosphatase (G6pc) (Fernandez-Larrea et al., 2007). Procyanidins might also act peripherally, since GSPE increases the mRNA expression of glucose transporter type 4 (Glut4) in white adipose tissue, although no significant changes in muscle of Glut4 or hexokinase 2 (Hk2) have been found (Fernandez-Larrea et al., 2007).

Finally, Li et al. tested the same dosage of GSPE (250 mg/kg), but for 24 weeks of treatment (Li et al., 2008). They found neither a change in glycemia or in HbA1c, but they found a clear improvement in body weight gain and several other disturbed aspects linked to diabetes. In fact, almost all the previously mentioned studies found improvements in inflammatory status (Lee et al., 2007), antioxidant effects (in almost all the studies), and improvements in diabetic nephropathy (Li et al., 2008).

Taken together these studies suggest that procyanidins have a short-lived insulin-mimetic effect on internal targets of the organism (Al-Awwadi et al., 2004; Maritim et al., 2003; Pinent et al., 2004) and are useful in improving the general situation of entire organism (Lee et al., 2008) most likely due to oligomeric forms (Lee et al., 2007) and to a forced acute dose (almost all summarized works in Table 5.1). What remains unclear is the ability of procyanidins to ameliorate the actual cause of type I diabetes, that is, to restore pancreas functionality {reviewed in (Pinent et al., 2008)}. Of the studies mentioned above, only (Al-Awwadi et al., 2004) analyzed plasma insulin, finding that procyanidins had no significant effect on it. This, in conjunction with the weak procyanidin effects found in other studies, does not support such amelioration of pancreas function. On the other hand, El-Alfy et al. used a different model for type I diabetes, alloxan-induced diabetes, and found a reduction in glycemia due to procyanidins associated with an increase in insulinemia, but only after 72 hr of GSPE (50 and 100 mg/kg bw, daily) treatment (El-Alfy et al., 2005). The reasons for the difference between the results of this study and those of other studies may include the time that the procyanidins were administered (El-Alfy et al. are the only ones who gave the first dose prior to drug treatment) but also to the different mechanisms used by streptozotocin and alloxan to impede pancreas functionality. Both diabetogens lead to β-cell loss, but while in alloxan-induced diabetes reactive oxygen mediates the selective necrosis of β-cells, in STZ-diabetes DNA alkylation might primarily mediate its toxic action, although production of reactive oxygen species may also be involved (Lenzen, 2008). Therefore, the antioxidant effects of procyanidins on the pancreas cannot be ruled out as a parallel mechanism for ameliorating type I diabetes in some situations.

Procyanidin Bioactivity in Healthy Animals (Normoinsulinemic)

The ability of procyanidins to mimic insulin effects in insulin sensitive targets when insulin is scarce or absent has been proved *in vivo* and shown as well *in vitro* (see paragraph of mechanisms of action). However, such effects are not so clear when there is a normal amount of insulin in the body.

Table 5.2 summarizes the results obtained from procyanidin treatments in healthy animals. Of all of them, only (Al-Awwadi et al., 2004) describe a slight but significant

Table 5.2. Summary of published effects of oligomeric procyanidins on glucose homeostasis in healthy animal models.

Estimated dose	Published dose	Extract	Duration	Animal model	Glucose measurement	Glucose effects	References
	10 mg/kg bw; daily; intra-peritoneally	Pycnogenol, a standard-ized extract from French maritime pine bark	14 days	Female Sprague-Dawley rat	Day 15; cardiac puncture; halothane anaesthesia	No change on serum glucose	Maritim et al., 2003
	20 mg/kg daily	provinols	8 weeks	Zucker lean rat	Not specified	No change glucose nor fructosamine	Agouni et al., 2009
19.5 mg/kg bw in sunflower oil		GSPE	4 weeks	Zucker lean rat	Not specified	No change glucose and slight decrease on body weight gain	unpublished results
Around 40 and 70 mg/kg bw in the pellet	Diet containing 0.5% or 1% CLP (2.2% catechin+ epicatechin; 1.69% B2; 2.37% C1; 2.01% cinnam-tannin A2)	Proanthocyanidins from cacao liquor	3 weeks	mice	Postprandial	No change	Tomaru et al., 2007
	200 mg “gallic eq”/kg; single dose at 9 a.m.	Red wine extract	6 weeks	Male Wistar rat	Monthly simultaneous to the procyanidins dose. End experiment and OGTT, 24 hr after last dose	Reduced body growth and blood glucose slight but significant	Al-Awwadi et al., 2004
250 mg/kg bw forced orally		GSPE	1 dose	Male Wistar rat	5 after dose	No change	Fernandez-Larrea et al.; 2007
	250 mg/kg bw forced orally	GSPE	24 weeks	Male Wistar rat	After overnight fasting	No effects in blood glucose	Li et al., 2009
1000 mg/kg bw forced orally		GSPE	1 dose	Male Wistar rat	4 after dose	No change	unpublished results

positive effect of procyanidins on plasma glucose. They used male Wistar rats as a healthy model simultaneously with the abovementioned type I diabetic animals and they describe similar but milder procyanidin effects in the normal rats, which they explain as due to a reduction in food intake and/or absorption. Meanwhile, a study conducted by our group in which we administered an acute dose of GSPE similar to that used by Al-Awwadi to male Wistar rats for five hours, does not support this effect (Fernandez-Larrea et al., 2007). We found the same changes in the gene expression of liver Gck and Gkrp as those observed in diabetic animals, but on the other hand we observed an up-regulation of G6pc gene expression, and remarkably plasma glucose was not modified. A higher acute dose (1g/kg bw) did not have any effect on plasma glucose either (unpublished results). Agouni et al. report no changes in plasma glucose or in body weight after treating lean male Zucker rats with a much lower dose (20 mg/kg) of provinols (Agouni et al., 2009). Esteve et al. (personal communication) conducted a similar study with GSPE in lean Zucker rats using a similar dosage of procyanidins, and did not find any effects on glycemia, although a slight limitation in weight gain was noted. Higher doses were assayed in mice by (Tomaru et al., 2007), but procyanidins were not found to affect glycemia.

If inhibition of glucose absorption was the main mechanism explaining the glucose-lowering effects observed in diabetic animals, similar effects would be expected in non-diabetic animals. However, most of the experiments shown here do not confirm that expectation. Therefore, all these data lead to the conclusion that when there is a lack of insulin, procyanidins may act as insulin-mimetic agents affecting some insulin targets. But under normal insulinemia, the results presented here suggest no clear effect of procyanidins on whole glucose homeostasis, probably due to the fact that insulin is more effective in terms of physiological effects.

Procyanidin Bioactivity in Insulin Resistant States

Procyanidins, therefore, do not seem to have a clear effect in normal conditions, but their interest as bioactive compounds lies in their beneficial effects against glucose homeostasis deregulation. Animal models for disturbed glucose homeostasis can be achieved by means of several different approaches, such as special diets. Table 5.3 summarizes studies in which procyanidins were given to animals simultaneously with the induction of insulin resistance by means of different diets. The first three summarized assays (Al-Awwadi et al., 2005; Tsai et al., 2008; Yokozawa et al., 2008) analyze the effects of procyanidins administered by oral gavage on fructose-induced insulin-resistant animals. Even though these studies differ in terms of dosages, treatment times, and the procyanidin extracts, all of them found an improvement in hyperglycemia as a result of procyanidin treatment. On the other hand, other studies summarized here found that procyanidins had little effect on hyperglycemia. We assayed the preventive role of a final dose of approximately 20 mg GSPE/kg bw given concurrently with two similar medium-fat diets (31.8% fat and 45% fat) in lean male Zucker rats and female Wistar rats, respectively, for a period of 19 weeks. In these studies, the GSPE was included in the animal feed (pellet). Male Zucker rats showed a slight improvement in glycemia along with improved adiponectinemia (Terra et al., 2009). In contrast, in female Wistar rats glycemia tended to increase and insulinemia decrease as a result of GSPE treatment, with no changes in adiponectinemia (Terra et al.). Since the GSPE extract used was the same in both studies, this different response could be partially explained by sex differences. In fact, female rats are more sensitive to high-fat treatments than males (Ribot et al., 2008) and this could mask the effects of GSPE, to which females would be more resistant. Glycemia was also not affected in a parallel group of female Wistar rats treated with a higher dosage of GSPE (35.8 mg/kg bw) included in a 60% fat diet, although increased insulinemia due to GSPE treatment was documented. Adiponectinemia also remained unchanged, despite an increase in its gene expression in mesenteric adipose tissue (Terra et al.). The differences in the efficacy of procyanidin extracts in different insulin-resistance induction models may be explained by the origin of the disturbances in each model. Fructose-fed disruption begins with the lack of control of hepatic lipogenesis (Li et al., 2006; Tran et al., 2009), which procyanidins have been shown to repress (Baiges et al., 2010). This effect, together with the fact that GSPE limits VLDL assembly (Bladé et al., 2010) would inhibit the induction of pain in the animal, limiting the development of insulin resistance in this model. Meanwhile, in high-fat models hepatic lipogenesis and VLDL assembly are not the key points, leading to metabolic disruption (Storlien et al., 2000; Varga et al., 2009), so the effects of procyanidins on lipid metabolism would not directly lead to the amelioration of glucose homeostasis.

On the other hand, Decorde et al. (Table 5.3) show that a GSPE dose similar to those previously discussed administered to hamsters for 12 weeks can reduce hyperglycemia in high-fat (44.4%) treated animals (Decorde et al., 2009). Again the differential effects of procyanidins on high-fat treated animals could be due to their effects on lipid metabolism, considering that hamsters have a very different lipid metabolism. But, it is also important to bear in mind that in this study GSPE was not included in the animals' feed pellets, but was administered daily by means of oral gavage. This suggests that procyanidins could also partially prevent high-fat diet-induced hyperglycemia, but only when they are acutely administered. (Serra et al., 2010) claim that plasma procyanidin levels depend on the vehicle. GSPE administered orally to rats along with carbohydrate enriched food led to lower plasma levels of oligomeric procyanidins than when GSPE was dissolved in water. And, it is important to note that higher amounts of procyanidins reach the circulation when they are administered with a daily bolus than if they are included in feed pellets, which the animals consume in small amounts throughout the day. Because this daily oral gavage was also used in the fructose-fed models in which GSPE showed positive effects, the mode of administration seems to be important, suggesting that a single dose of GSPE may be effective in lowering glycemia. However, this single dose might not always be effective; Zhang et al. also administered procyanidins once daily by oral gavage (80 mg/kg bw for six weeks) and found no amelioration of plasma glucose levels. This study was performed in mice, so the use of a different species could also explain the different results compared to the abovementioned studies (Zhang et al., 2009).

Another situation in which procyanidins might have beneficial effects is their possible role in improving or correcting an already disturbed glucose homeostasis. Table 5.4 compiles the studies that analyze these effects in different animal models. The cafeteria diet is a successful model for inducing a clear dysfunction in the metabolism

Table 5.3. Summary of preventive effects on different models of insulin resistance.

Estimated dose	Published dose	Extract	Duration	Animal model	Glucose effects	References
	21.42 mg/kg oral gavage	Vitaflavan vs. antocianins vs. procyanidins	6 weeks	Sprague-Dawley simultaneous to 66% fructose	Decrease glucose. no changes insulin	Al-Awwadi et al., 2005
	10 or 20 mg/kg bw; oral gavage	Grape seed proanthocyanidin (gravinol)	2 weeks	Male Wistar simultaneous to 65% fructose	Lowers glucose and bw	Yokozawa et al., 2008
	125 mg/bw and 250 mg/bw; oral gavage at 6 p.m.	Proanthocyanidin-rich extract from longan flowers	14 weeks	Male Sprague-Dawley simultaneous to 60% fructose	Lowers glucose, insulin. Improves Irs1 and Glut4 protein in WAT mes	Tsai et al., 2008
21.7 mg GSPE/kg bw included on the pellet	Nearly 7 mg/animal/day	GSPE	19 weeks	Male Zucker lean simultaneous to 32% fat diet	Slight decrease on glucose, nor insulin, nor bw, adiponectin increased	Terra et al., 2009
20.15 mg GSPE/kg bw included on the pellet		GSPE	19 weeks	Female Wistar simultaneous to 45% fat diet	No changes	submitted to ICP
35.8 mg GSPE/kg bw included on the pellet		GSPE	19 weeks	Female Wistar simultaneous to 60% fat diet	No changes	submitted to ICP
35.7 mg/kg oral gavage	7.14mg/kg bw	Chardonnay GSPE	12 weeks	Male Hamster Simultaneous to 44.36% fat diet	Decrease glucose. No changes insulin and increase in adiponectin. Lowers bw	Decorde et al., 2009

Table 5.4. Summary of corrective effects on glucose homeostasis in animal models.

Estimated dose	Published dose	Extract	Duration	Animal model	Glucose effects	References
	25 or 50 mg/kg bw voluntarily in the morning	GSPE	10 days	Female Wistar rat kafeteria treated	No change glucose, slightly higher insulin	Montagut, Blade, et al., 2009
	25 or 50 mg/kg bw voluntarily in the morning	GSPE	4 weeks	Female Wistar rat kafeteria treated	Slight decrease on glucose simultaneous to insulin	Montagut, Blade, et al., 2009
Around 40 and 70 mg/kg bw in the pellet	Diet containing 0.5% or 1% CLP (2.2% catechin+ epicatechin; 1.69% B2; 2.37% C1; 2.01% cinnamtannin A2)	Proanthocyanidins from cacao liquor	3 weeks	Db/db mouse	Decreased glucose, dose-dependent and fructosamine	Tomaru et al., 2007
	10 mg/kg bw orally forced	Oligomeric procyanidins (MPD: 3.3)/polymeric. All from persimmon peel	6 weeks	Db/db mouse	Decreased serum glucose, and glicosilated protein	Lee et al., 2008
	20 mg/kg	provinols	8 weeks	Zucker fa/fa rat	Decreased serum glucose, fructosamine	Agouni et al., 2009
19.5 mg/kg bw vehicle: sunflower oil, forced orally,		GSPE	4 weeks	Male Zucker fa/fa rat	No changes	Dr. Esteve personal communication

Note: This table summarizes the studies in animals with genetically-induced insulin resistance and studies in which insulin resistance was induced using different diets, and procyanidins were administered afterwards, once the animals were already damaged.

of rats (Petry et al., 2001). In this model, an acute daily dose of 25 mg GSPE/kg bw dissolved in semi-skimmed milk for 30 days improved the glucose homeostasis of diet-induced (high fat, high sucrose) insulin resistant animals. However, the effects of GSPE are dependent on the dosage and time of treatment, since neither shorter treatments nor a higher dosage were found to have a clear positive effect on glucose homeostasis (Montagut, Blade, et al., 2009).

Glucose homeostasis disruption can also be studied using genetically altered models. Table 5.4 also contains experiments performed with such models, in which the effects of procyanidin extracts are more difficult to understand. A similar dose (19.5 mg/kg bw) of the same GSPE as that previously shown to be effective on cafeteria-diet fed animals, daily force-fed (sunflower oil as vehicle) to male Zucker fa/fa rats did not improve glycemia at all (Dr. Montserrat Esteve, personal communication). In contrast, orally force-fed provinols at 20 mg/kg bw have been found to improve glycemia in the same animal model (no information concerning animal sex is provided in this case) (Agouni et al., 2009). There are numerous differences between these studies, such as the vehicle, the duration of the treatment, and the flavonoid extract. The main difference between the two extracts is their monomeric components: anthocyanidins in provinols and flavanols in GSPE. One might speculate that the different compositions are the source of the different responses. In fact, in the abovementioned study by (Al-Awwadi et al., 2005), slight differences were observed between extracts rich in either anthocyanidins or procyanidins, although in that case the procyanidin-enriched extract was the most effective at ameliorating insulin sensitivity. Therefore, the conclusion cannot be drawn that a particular type of flavanol is responsible for the effects of the extracts.

In db/db mice studies, GSPE improved glucose homeostasis at very low doses: approximately 10 mg/kg bw daily forced (Lee et al., 2008) and 40 or 70 mg/kg bw in feed pellets (Tomaru et al., 2007). The same genetic alteration is found in db/db mice as in obese Zucker fa/fa rats (Tartaglia, 1997), but the genetic background of each species gives rise to metabolic differences in their disturbed metabolic profiles, such as different β-cell characteristics. According to (Shafrir et al., 1999), fa/fa rats have "sturdy" β-cells that maintain a robust, lifelong insulin secreting capacity. Meanwhile, db/db mice are characterized by "brittle" β-cells that allow only transient insulin hypersecretion with short-term obesity. Little research has been done into the effects of procyanidins on β-cells (Pinent et al., 2008), and preliminary results from our research group indicate that β-cells are a target for procyanidins (Castell et al., 2009). So, differences in β-cell characteristics may be one explanation for the different responses to GSPE in fa/fa rats and db/db mice, although more work needs to be done to get a more complete picture of its effect on genetically altered animal models.

In conclusion, GSPE might improve a slightly disrupted homeostatic situation, but such effects are highly dependent on the quantity of procyanidins that the animals receive, including the daily dose, which in turn depends on the method and period of administration.

Mechanisms Used by Procyanidins to Improve Glucose Homeostasis

The collected data show that procyanidins might be able to ameliorate glucose homeostasis disturbances in specific situations. The mechanisms used by procyanidins to exert such effects are diverse, as summarized here.

Firstly, after the ingestion of food, limiting post-prandial glycemia increases would be beneficial against insulin resistance. In this respect, inhibitors of the enzymes that participate in carbohydrate digestion would have a positive effect. In fact, procyanidins derived from different plant origins can inhibit α-glucosidases and α-amylases (Lee, Cho, et al., 2007; Loo and Huang, 2007; Schäfer et al., 2007; Yuste et al., 1992). Such effects are observed mainly in polymeric forms and are not reproduced with monomers (Loo and Huang, 2007; Schäfer et al., 2007) and involve a mechanism that is not fully elucidated. Most of these studies analyze the inhibitory capacity *in vitro*, but little evidence has been found that reinforces their actual role *in vivo*. Another possible mechanism for reducing the level of glucose entering the organism is the inhibition of its absorption via inhibiting the transporters involved in glucose uptake to the intestine. Some monomeric flavonoids have been shown to inhibit the sodium-coupled glucose transporter (Sglt1) (Kottra and Daniel, 2007) and thus reducing glucose uptake. Some monomeric flavan-3-ols have also been found to inhibit the response of Sglt1 (Hossain et al., 2002; Kobayashi et al., 2000). Glucose transporter type 2 (Glut2)-mediated glucose uptake is also inhibited by some flavonoids such as quercetin, but not by the procyanidin monomers such as catechin or epicatechin (Kwon et al., 2007). The effects of oligomeric procyanidins, however, have not been explored. These procyanidin mechanisms of reducing blood glycemia should be effective in any condition of glucose homeostasis, but that is not always the case—it seems that this mechanism is not always active, especially in healthy rats. And in long-term treatments of diet-induced insulin-resistance it seems that when procyanidins are administered in a daily bolus they are more effective than when given with food, which indicates the present mechanisms of action other than those inhibiting carbohydrate digestion and/or glucose absorption.

These other mechanisms would involve targeting the liver. The liver plays a central role in maintaining glucose homeostasis, and the accumulation of hepatic lipids may be an important factor contributing to insulin resistance. Several flavonoids have been shown to increase the gene expression of Gck and reduce that of G6pc (AePark et al., 2006; Jung et al., 2006; Wolfram et al., 2006). Such effects have also been shown for procyanidins in some conditions, for example in STZ-treated rats where procyanidins ameliorated glucose homeostasis (Fernandez-Larrea et al., 2007; Pinent et al., 2004). Gck activity has also been increased by grape seed procyanidins in high-fat diet treated rats (Zhang et al., 2009), although not by enough to ameliorate glycemia. And, in cafeteria-diet induced diabetic animals the amelioration of glucose homeostasis by 25 mg/kg bw of GSPE was not accompanied by such gene expression changes (Montagut, Blade, et al., 2009). Therefore, when there is a lack of insulin, procyanidins might act as an insulin-mimetic, acting on the insulin targets Gck and G6pc, but chronically administered in insulin-resistance conditions, the effects of procyanidins are not so clear. So, control of glucose input/output by the liver could be one of the mechanisms used by procyanidins to improve glycemia, but would not be enough to fully explain their effects, and depend on the insulinemia status of the organism.

On the other hand, 50 mg/kg bw of GSPE administered to cafeteria-diet fed animals (Montagut, Blade, et al., 2009) did target the liver limiting lipogenesis. So the reduction in hepatic lipogenesis might be a mechanism for improving insulin resistance in models in which the increase of lipid synthesis in the liver is a key factor in the development of diabetes, but might not be so important in models in which insulin resistance is mainly caused by other factors.

The effects of procyanidins might also be mediated peripherally. *In vitro* studies suggest that procyanidins have insulin-mimetic effects in adipose tissue and muscle (Blay, 2003; Li et al., 2007; Montagut et al., 2009; Pinent, Bladé, Salvadó, Blay, et al., 2006; Wolfram et al., 2006). We have shown that the oligomeric procyanidins of GSPE activate the insulin receptor by interacting with and inducing its phosphorylation and that this interaction leads to increased glucose uptake mediated by Akt. However, GSPE phosphorylates proteins of the insulin signaling pathway differently than insulin does, as Akt, p44/42 and p38 MAPKs are the key points for GSPE-activated signaling mechanisms, suggesting alternative pathways for procyanidins to be effective in insulin-resistant situations (Pinent et al., 2006).

In the type I diabetes model, we found increased mRNA expression of Glut4 in white adipose tissue, which supports the hypothesis that when there is a lack of insulin procyanidins might act as insulin-mimetics. The remainder of the studies listed did not analyze the adipose or the muscle tissue after procyanidin treatment. On the other hand, long-term GSPE treatment in insulin-resistant rats down-regulated the gene expression of Glut4, concomitantly with a down-regulation of several adipose cell markers (Montagut, Blade, et al., 2009). Actually *in vitro* procyanidins have been shown to inhibit adipogenesis, pointing to the modulation of adipose tissue depots as a mechanism for procyanidins to ameliorate glucose homeostasis (Pinent et al., 2005), though the effect of procyanidins on adipose tissue depots is beyond the scope of this review. Considering the studies that analyze the effects on glucose homeostasis, in the study of grape seed procyanidins and a cafeteria diet, these effects on adipose markers cannot be linked to the amelioration of glucose homeostasis, since similar changes were also found in the model where no amelioration occurred. The key controller of adipogenesis, Pparg, was also down-regulated in epididymal fat in obese Zucker rats treated with GSPE (Montagut et al., 2007), but no changes in insulin resistance were observed. In our opinion, the study periods of this research were not long enough to observe significant changes in adipose tissue depots. Both Yokozawa and Decorde found reduced adipose tissue and body weights related to the reduction of glycemia in their studies, although no further analysis of the adipose tissue was performed (Decorde et al., 2009; Yokozawa et al., 2008). Tsai et al. found increases in Irs1 and Glut4 proteins, which are related to the amelioration of insulin resistance (Tsai et al., 2008). Therefore, procyanidins might target adipose tissue in a few different ways: acutely, by acting as an insulin-mimetic; and chronically, by exerting changes in adiposity which in the long term would be beneficial against insulin resistance, and by improving glucose uptake by acting at the level of the Glut4 protein, or by means of other mechanisms not fully understood.

Finally, it is also possible that chronic procyanidin treatment targets the pancreas, the organ responsible for insulin secretion after glucose intake. Our group's preliminary results show that the pancreas might be a target tissue for procyanidins. In fact, other flavonoids have been shown to target the pancreas, although their exact role remains unclear (Pinent et al., 2008). This possibility therefore deserves further study.

Human Studies

Using the same approach as in the animal analyses, Table 5.5 summarizes the human studies of the effects of procyanidins on different metabolic situations. It is important to note that most of these studies on the effects of procyanidins were conducted with small samples, which diminishes the value of their results as representative in the general population.

The first part of the Table 5.5 shows healthy, normoinsulinemic people. Only 500 mg of cocoa polyphenol intake for two weeks resulted in improvements as measured by an OGTT (Grassi, Lippi, et al., 2005), suggesting improved insulin sensitivity. In contrast with these results, no effects were found in the studies conducted by (Baba et al., 2007) and (Zern et al., 2005), which used a larger sample size and higher doses of other extracts. Therefore, either these doses were too high or the treatments too long or these results are consistent with the animal results in normoglycemic situations, in which procyanidin treatment was ineffective.

The next studies do not refer to any detailed alterations in glucose homeostasis, but were conducted using subjects that suffer from some pathology. Most of the studies did not find any chronic effect of ingested procyanidins on glycemia or related parameters (glycosylated hemoglobin, insulin). Only Grassi et al. found a clear improvement in insulin sensitivity parameters, both in hypertensive subjects (Grassi et al., 2005), and hypertensive subjects with IGT (Grassi et al., 2008). All these studies refer to an acute dose of 500 mg of polyphenol in the form of a 100 g chocolate bar for a period of two weeks. A parallel study by Muniyappa (Muniyappa et al., 2008) with the same number of subjects (20 hypertensives) also for two weeks, but with two daily doses of cocoa extract (451 mg polyphenols/dose) dissolved in water, did not show any improvement in insulin sensitivity parameters.

The last studies were conducted using samples of subjects with type 2 diabetes. Two studies by the same author using the same extract (pycnogenol) show positive results. Liu et al. described a dose-response effect with dosages of up to 200 mg for three weeks (Liu et al., 2004). Banini et al. also found improved glycemic control with 150 ml of muscadine grape products over a period of four weeks. On the other hand, again with a much higher dose (6.8 mg GSPE/kg bw in tablets) for four weeks, no effect on glucose metabolism was documented.

None of the human studies report information about direct mechanisms that would explain the bioactivity of the compounds. Only some studies postulate a relationship between the effects and the clearly identified ability of procyanidins to improve

Table 5.5. Summary of studies on procyanidin effects on human studies.

Daily dose	Duration	Beneficial effect	n particpants/ treatment	Characteristics of participants	References
250 ml/day Sicilian red wine during meals	4 weeks	No changes glucose	24	healthy people; both sexes	Avellone et al., 2005
50 ml Promeganate Juice (PJ)/d (1.5 mmol total polyphenols). The soluble polyphenol content in PJ varies (0.2–1.0%); includes mainly anthocyanins, catechins, ellagic tannins, and gallic and ellagic acids	2 weeks	No changes glucose	-13	healthy, non-smoking men, aged 20–35 years	Aviram et al., 2000
Increasing doses of PJ (20–80 ml/d, equivalent to 0.54–2.16 mmol total polyphenols/d)	10 weeks		-3		
Six pairs of foods +/– chocolate (do not specify the amount of chocolate!!)	One dose	Chocolate flavoured version 28% greater insulinemia than the alternate product	10	Healthy people, both sexes	Brand-Miller et al., 2003
500 mg polyphenols as 100 g dark chocolate bar	2 weeks	Improved Homa-IR, Quicki, and lower glucose and insulin after OGTT	15	Healthy people, both sexes	Grassi, Lippi, et al., 2005
26 g cocoa powder and 12 g sugar/day; twice each day: before noon and during the afternoon.	12 weeks	No significant differences	28	Subjects fasted for 12 hr, and then blood samples were collected from the intermediate cubital	Baba et al., 2005
(units/100 g cocoa powder) 377 mg epicatechin, 135 mg catechin, 158 mg procyanidin B2, 96.1 mg procyanidin C1, 2192 mg theobromine, and 470 mg caffeine					
36 g of a lyophilized grape powder (LGP) or a placebo	4 weeks	Glucose concentrations were not affected	24–20	pre- and post-menopausal women	Zern et al., 2005
LGP total phenols, 5.8 g/kg; flavans, 4.1 g/kg; anthocyanins, 0.77 g/kg; quercetin, 102 mol/kg; myricetin, 8 mol/kg; kaempferol, 11 mol/kg; and resveratrol, 7 mol/kg					

Table 5.5. *(Continued)*

Daily dose	Duration	Beneficial effect	n particpants/ treatment	Characteristics of participants	References
100 ml red grape juice (RGJ)/d	2 weeks	Plasma concentrations of glucose, did not change	26 patients/15 healthy	patients receiving hemodialysis and healthy subjects	Castilla et al., 2006
2* 180 mg polyphenols	4 weeks	no changes on HbA1c	20	Normotensive with increased cholesterol	Allen et al., 2008
1400 mg polyphenols 1.06 g procyanidins (with other polyphenols, total 1.44)	16 weeks	No changes on glucose, nor glycosilated hemoglobin	21/13	Normocholesterolemic/hypercholesterolemic	Jiménez et al., 2008
900 mg flavanols	2 weeks	Improve insulin resistance	20	Essential hypertension	Muniyappa et al., 2008
500 mg polyphenols as 100 g dark chocolate bar	2 weeks	Improved Homa-IR, Quicki, and ISI (insulin resistance index)	20	Both sexes, hypertensives	Grassi et al., 2005
500 mg polyphenols as 100 g dark chocolate bar	2 weeks	Decreased insulin resistance and increased insulin sensitivity (ISI), and {beta}-cell function	19	Hypertensives with IGT	Grassi et al., 2008
190 ml red wine, 13% v/v ethanol, 5 days per week	10 weeks	No changes on glucose, nor insulin	20	Overweight women (BMI: 29.8 ± 2.2 kg/m^2)	Cordain et al., 2000
640 ml/day purple grape juice	8 weeks	No changes glucose, nor insulin	22	Coronary artery disease	Chou et al., 2001
0.7±1.2 ml/kg*d of purple grape juice	2 weeks	Baseline: insulin 7.52 ± 6.03 μIU/ml. After GJ: insulin 20.49±12.55 μIU/ml (*P*=0.004).	15	Mixed adults with angiographically documented CAD	Stein et al., 1999
100 mg pycnogenol (1.45 mg/kg bw)	12 weeks	Decrease glucose, initial decrase of HbA1	34	T2DM with treatment	Liu, Wei, et al., 2004
50, 100, 200, 300 mg pycnogenol (bw??)	3 weeks each dose, final treatment 12 weeks?	Decrease glucose dose-dependent until 200 mg. decease HbA1c no changes insulin	40	T2DM	Liu, Zhou, and Rohdewald, 2004
6.8 mg GSPE/kg bw tablets	4 weeks	No improved glucose metabolism. No change HOMA-IR. Decreased frutosamine	32	T2DM with treatment	Kar et al., 2009
150 ml grape wine or dealcholized	4 weeks	Improved glicemic control	10	T2DM	Banini et al., 2004

inflammatory response (Grassi et al., 2005; Grassi, Lippi, et al., 2005; Grassi et al., 2008). So, from all these studies, the clearest conclusion that can be drawn is a rough estimation of the effective dose in humans: around 100–500 mg polyphenols daily, acutely administered for two to three weeks. It is important to bear in mind that this dosage is 10 times higher than estimated normal intake (Grassi et al., 2005; Grassi, Lippi, et al., 2005; Grassi et al., 2008).

From all the information published in procyanidin studies on glucose homeostasis regulation, we can conclude that procyanidins are more effective when administered in one acute load, and that they act as insulin-mimetics when there is no (or little) insulin present, probably targeting the liver and peripheral tissues. In a glucose-disturbed metabolism, procyanidins are useful because they prevent the induction of damage (fructose-fed models) and/or because they may use alternative targets to exert their insulin-mimetic effects. The validity of these mechanisms in humans cannot be deduced from current research, because such studies are scarce and tend to be conducted with limited sample sizes.

LIST OF ABBREVIATIONS

bw: body weight
GSPE: Grape Seed Procyanidin Extract
Gck: Glucokinase
Gckr: Glucokinase-regulatory protein
G6pc: glucose-6-phosphatase
Glut4: Glucose transporter type 4
Hk2: hexokinase 2
STZ: streptozotocin
MPD: mean polymerization degree

SUPPLEMENTARY DATA

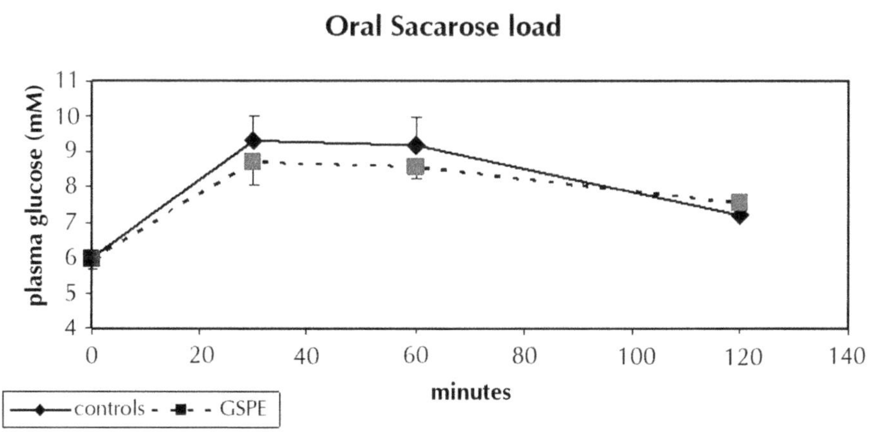

Figure 1. Oral sucrose load on healthy rats, simultaneously to a high GSPE oral acute dose.

ACKNOWLEDGMENTS

This study was supported by grant numbers AGL2008-01310 and AGL2008-00387 from the Spanish Government. Gemma Montagut received a fellowship from Universitat Rovira i Virgili in Tarragona. Lidia Cedó received a fellowship from the Generalitat de Catalunya. We would also like to thank the language service of the Universitat Rovira i Virgili for revising this manuscript.

There are no financial or commercial conflicts of interest.

KEYWORDS

- **Diabetes**
- **Glucose**
- **Human studies**
- **Insulin resistance**
- **Procyanidin**
- **Rat**

Chapter 6

Advances and Applications of Galactosidases in Food Industry

Carlos Ricardo Soccol, Adenise Lorenci Woiciechowski, Michele Rigon Spier, Adriane Bianchi Pedroni Medeiros, and Akbar K. Haghi

INTRODUCTION

Some food products have nutritional properties, but its consumption is somewhat limited in part of the population due to some digestion problems or intolerances it provokes. This chapter focuses on the importance of galactosidases to improve health and quality of foods containing, reducing the levels of indigestible oligosaccharides and disaccharides. Several studies have shown the production and the application of α and b-galactosidases to reduce undigested components, naturally present in foods to minimize digestion problems. α-galactosidases is a glycoside hydrolase enzyme that hydrolyses the terminal α-galactosyl from glycolipids, glycoproteins, and non-digestible oligosaccharides such as galactooligosaccahrides, raffinose and stachyose, present in soy-derived products, limiting human consumption of these products. This enzyme is not synthesized by mammals, including man which is necessary for the hydrolysis of these sugars. β-galactosidases is another group of galactosidases also important in food processing mainly in dairy-products processing to achieved lower lactose content to people who has lactose intolerance. β-galactosidases is a hydrolase enzyme that catalyzes the hydrolysis of β-galactosides into monosaccharides, acting on lactosylceramides, lactose, and various glycoproteins (Dorland's Illustrated Medical Dictionary, Retrieved October 22, 2006), lactase being a sub-class of β-galactosidase. Researches in galactosidases' production and purification are being done to apply these enzymes in new dairy-products. The development of new food products containing healthy benefits and which minimized digestive problems are world-wide consumption trends.

Many vegetables and dairy products important to the human dietary have nutritional properties, but contain considerable levels of undesirable components such as tannins, lectins, protease, and amylase inhibitors and the presence of oligosaccharides due to some digestion problems it provokes. Lactose, raffinose, stachyose, mellibiose, and verbascose are examples which are not hydrolyzed in the small intestine due to a lack of the specific hydrolase enzyme, but are passed into the lower gut, where they are fermented by bacterial colonies which are able to ferment these sugars with the liberation of a considerable amount of gas (containing mainly carbon dioxide, hydrogen, and methane) often causing flatulence (Iyer et al., 1980; Vidal-Valverde et al., 1993). Other symptoms that can appear are nausea, cramps, abdominal pain (Cristofaro et al.,

1974), headache, mental and concentration disturbance (Sannie et al., 1997). For that reasons, α and β-galactosides are considered arduous factors (Musquiz et al., 1999). On the contrary, there are a lot of evidences to support the beneficial effect of the presence in the diet of these sugars and particularly of oligosaccharides derived from inulin and commonly named oligofructose (Tomomatsu, 1994).

The main goal of this chapter is focuses on importance of galactosidases at the degradation of raffinose family of oligossacharides present in vegetable and food products, as far as the importance to hydrolyse lactose of dairy products enabling the use of many products by persons who have any kind of intolerance to these oligosaccharides and review the enzymes which hydrolysis these antinutritional factors, found in vegetable or dairy sources or produced by bioprocesses.

OLIGOSACCHARIDES

Oligosaccharides consists of 2–10 monosaccharide residues linked by glycosidic linkages which can be liberated by depolymerization (Khadem, 1988; Pazur, 1970). Although the oligossacarides brings beneficial effects such as prebiotics (frutoligossacharides), and the potential of carbohydrates in the maintenance of health and as therapeutic agents has been realized (McAuliffe and Hindsgaul, 1997; Sharon and Lis, 1993), some of them are known to have antinutritional effects, as the raffinose family of oligossacharides. The α-galactosidases enzyme group act in hydrolysis these oligossacharides. Lactose, a disaccharide, can also be named as lactase, an enzyme of the β-galactosidases group that acts to hydrolyse this sugar.

RAFFINOSE FAMILY OF OLIGOSACCHARIDES

The raffinose family of oligosaccharides or α-galactosides belongs to compounds widespread in the higher plants. Large amounts of them occur in plants where they constitute the main fraction of water soluble carbohydrates. These oligosaccharides are very important components of the carbohydrates reserve in the vegetative storage organs and seeds of many plants and they play various physiological functions (Musquiz et al., 1999) comprise 30–80% of the total soluble sugars.

The raffinose family of oligosaccharides has a more definitive antinutritional effect (Deshpande, 2002). The absence of the α-galactosidase enzyme in the low gastrointestinal tract (responsible for the hydrolysis of α-1, 6 galactosides linkages) and its accumulation in the large intestine, results in the fermentation by anaerobic bacteria (Fleming, 1981). It is the behavior that produces the flatulence for which the consumption of this oligossacharides is noted (De Lumen, 1992; Viana et al., 2005). The accumulation of flatus in the intestinal tract results in discomfort, abdominal rumblings, cramps, pain, and diarrhoea and is characterized by the production of hydrogen, carbon dioxide, and small amounts of methane gas. Sugars of the raffinose family are believed to be largely responsible for the often reported problem of flatulence after consumption of diets containing beans and other legumes.

The raffinose family of oligosaccharides is represented by the predominant oligossacharide stachyose (tetrassacharide), followed by raffinose (trissacharide) and verbascose (pentassacharide), depending on the type of grain (FAO, 1998; Petterson and Mackintosh, 1994). On the oligosaccharide structure, sucrose may contain one or more α-D-galactopyranosyl (α-D-Gal) groups, in which the α -galactosyl units are always to the left of sucrose, connected to the molecule of glucose (Silva et al., 1992). The structural relationship of these sugars is showed in Fig. 6.1.

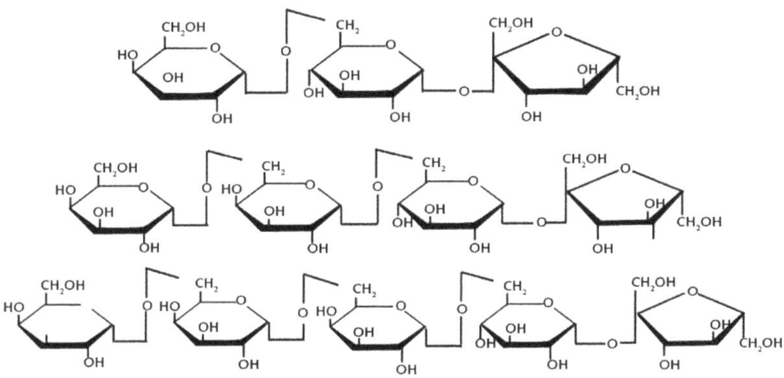

Figure 6.1. Structure of raffinose (A), stachyose (B), and verbascose (C).

Raffinose

Raffinose is a trisaccharide composed of glucose, fructose, and galactose and is commonly found in various vegetable such as grains (beans and soybeans), and other vegetable such as cabbage, broccoli, asparagus, and brussels sprouts. Raffinose can be hydrolyzed to D-galactose and sucrose by the enzyme α-galactosidase, an enzyme not found in the human digestive tract. α-galactosidase also hydrolyzes other α-galactosides such as stachyose, verbascose, and galactinol, if present. The enzyme does not cleave β-linked galactose, as in lactose (Insel et al., 2011).

Stachyose

Stachyose is formed from four monosaccharide molecules—two galactoses, one glucose, and one fructose, which are readily broken down by intestinal bacteria, and are responsible for the familiar gaseous effects of foods, such as beans (Insel et al., 2011).

Verbascose

Verbascose is a pentasaccharide formed by three galactoses, one glucose, and one frutose, of the raffinose family of oligosaccharides occurs in many plant species, specially in seeds (Dey, 1985; Peterbauer et al., 2003). As its lower homologues raffinose and stachyose, verbascose may act as a protective agent during maturation drying, or as a carbon reserve for germination (Obendorf, 1997). Considerable variation exists in the content of verbascose in seeds, varying from 1.1 to 9.9% (of seed dry matter faba bean) (Kuo et al., 1988) (Horbowicz and Obendorf, 1994), achieving 3.1% in pea (*Pisum sativum*) (Jones et al., 1999). Due to their anti-nutritional characteristics, elimination of them from crop seeds remains an attractive breeding target (Hitz et al., 2002), although recent research also suggests that health benefits may be associated with the consumption of these oligossacharides (Aranda et al., 2000).

SOURCE OF OLIGOSACCHARIDES

Legumes constitute one of the richest and least expensive sources of proteins in the human diet in many parts of the world. Proteins are essential components of the diet, needed for survival of humans and animals. The basic function of proteins in nutrition is to supply adequate amounts of required amino acids. Soybean (*Glycine max*) is one such important plant in human and animal nutrition. About 60% of all processed food products include ingredients derived from soy. Soy products have superior nutritional characteristics in terms of high protein, amino acids contents and it is suitable as a lowcost substitute to dairymilk in developing countries and as a nutritive supplement for the lactose intolerant population (Biranjan et al., 1983). Furthermore, it contains a large amount of galactooligosaccharides such as raffinose and stachyose that are indigestible by mammals and are flatulence-causing factors (Reddy et al., 1980). Flatulence is the common symptom associated with consumption of these oligosaccharides, but more serious accompanying consequences such as abdominal pain and diarrhoea overshadow its social implications (Salunkhe and Kadam, 1989).

Beans (*Phaseolus vulgaris* L.) are important sources of proteins, carbohydrates, fibers, vitamins, unsaturated fatty acids, and micronutrients (Deshpande, 1992). Brazil is the first world producer of the species *Phaseolus vulgaris* (Hungria et al., 2000). In 2007, the Brazilian cultivation of beans achieved 3.2 million of tons, being a basic food to the population. The consumption of cooked beans is 12.7 Kg/person/year in Brazil (Embrapa, 2008).

Bean grains contain 20–35% protein and 50–60% carbohydrates, but present antinutritional factors such as galactooligosaccharides (GO), mainly raffinose, stachyose and verbascose (Lyimo et al., 1992). The GO are heat resistant and remain in the end product (Jangchud and Bunnang, 2001). Due to the absence of α-galactosidase (EC 3.2.1.22) in the intestinal tract of humans, the GO are accumulated and fermented, producing a large amount of gases. The GO are considered inducers of gastrointestinal disorders, diarrhoea and flatulence, inhibiting nutrient uptake (De Lumen, 1992).

Raffinose negatively impacts sugarbeet (*Beta vulgaris* L.) industrial processing by decreasing extractable sucrose yield and altering sucrose crystal morphology which reduces filtration rates and slows processing. Although increased raffinose concentrations have been observed during cold storage, the physiological and biochemical mechanisms associated with raffinose accumulation in sugarbeet are poorly understood (Haagenson et al., 2008). Table 6.1, presents the oligosacharides content of some sources.

Table 6.1. Source of oligosacharides and their respective content.

Plant	Raffinose	Stachyose	Verbascose	Reference
Lupinus albus	19.0%	54.0%	10.8%	Musquiz et al. (1999)
Lupinus luteus	9.6%	40.4%	35.7%	
Glycine max	28.3%	36.7%	0	
Phaseolus vulgaris	6.4%	62.1%	0	
Medicago sativa	23.1%	22.6%	44.7%	
Trifolium repens	18.5%	60.5%	0	
Ornithopus sativus	22.6%	19.6%	49.7%	
Sugarbeet root	0.8–6.0 g/Kg	nd	nd	**Haagenson et al. (2008)**
Cowpea	1.19g/100g	3.88g/100g	nd	Onyenekwe et al. (2000)

STUDIES SHOWING OLIGOSACHARIDES EFFECTS

The ingestion of galactooligosaccharides showed an increase in the incidence of diarrhoea in mice. This fact was confirmed by Liying et al. (2003), when a diet supplemented with 2% purified stachyose increased the occurrence of diarrhoea in pigs when compared to control.

Martínez-Villaluenga et al. (2006) reported the evidence that raffinose family of oligossacharides have beneficial effects on the survival of these probiotic cultures in dairy products. As a result, such stored dairy products containing both probiotics and prebiotics have synergistic actions in the promotion of health.

An application of α-galactosidase produced by *Mortierella vinacea* was potentially useful for processing of food products containing raffinose oligosacharides and the studies proved that this enzymatic digestion of lentil seeds with this enzyme was the most effective method of removal of raffinose which induces flatulent phenomena in humans after ingestion of legumes (Miszkiewicz and Galas, 2000) (see Table 6.1).

Although many high-protein foods such as meats, dairy products and beans are quite difficult for some people to digest, tofu is soft and highly digestible food. It presents a digestion rate of 95%, it is certainly the most digestible of all natural soybean foods, and is much more digestible than cooked whole soybeans (68%) due to the reduced oligosaccharides content (Shurtleff and Aoyagi, 1999).

ENZYME TREATMENT IN OLIGOSACHARIDES REDUCTION

Raffinose oligosaccharides (ROs) make up a substantial part (40%) of the soluble sugars found in soybean seeds and are responsible for flatulence after the ingestion of soybean and other legumes. Consequently, soy-based foods would find a broader approval if the ROs were removed from soybean products or hydrolysed by α-galactosidases.

Khare et al. (1994) have successfully demonstrated the removal of significant amount of oligosaccharides by *Aspergillus oryzae* cells entrapped in agarose. Their results indicate a possibility of using whole cells for pre-treatment of soymilk to overcome the physiological problems of flatulence. They suggest constructing a fluidized bed reactor for continuous hydrolysis of RFOs in soymilk. The aim of that work was to investigate the applicability of the immobilized cells for the reduction of RFOs in soymilk in continuous mode using a fluidized bed reactor to solve the physiological problem of flatus in a systematic manner.

Some authors have studied conditions for the decomposition of raffinose family sugars present in vegetables sources. Although thermal treatment such as cooking and autoclaving may reduce these groups, only enzyme treatment showed better efficiency in the removal of the oligosaccharides. Table 6.2 presents some studies of treatment for oligosacharides reduction.

Table 6.2. Oligosacharides decomposition reported in literature.

Oligosacharide	Source	Conditions	Reduction (%)	Reference
Raffinose and stachyose	Lentil seed	α-galactosidases from *Mortierella vinacea* IBT-3; 15% w/v suspension of lentil flour, 0.85 U.g^{-1}, one hour, 50°C, pH 5.0	95% of stachy-ose 90% of raffinose	Miszkiewicz and Galas (2000)
Raffinose and stachyose	Cowpea flour	α-galactosidases from *A. niger*	93% of raffinose and 82% of stachyose	Somiari and Balogh (1992)
Raffinose and stachyose	Chickpea flour	α-galactosidases from *Cl. Cladosporides* (40°C, pH 5.0, 290 U/ml)	100%	Mansour and Khalil (1998)
Raffinose and stachyose	Chickpea flour	α-galactosidases from *A. oryzae* (50°C, pH 4.5, 210 U/ml)	100%	Mansour and Khalil (1998)
Raffinose and stachyose	Chickpea flour	α-galactosidases from *A. niger* (50°C, pH 5.0, 130 U/ml)	100%	Mansour and Khalil (1998)
Raffinose and stachyose	Soymilk	α-galactosidases from *Cl. cladosporides*	100%	Cruz and Park (1982)
Raffinose and stachyose	Cowpea	Cooking with potash leads	68.9% raffinose 61.34% stachy-ose	Onyenekwe et al. (2000)
Raffinose	Cowpea	Cooking in tap water	50%	Onyenekwe et al. (2000)
Stachyose	Cowpea	Cooking in tap water	<50%	Onyenekwe et al. (2000)
Raffinose and stachyose	Cowpea	Processing into bean cake	52% raffinose 52% stachyose	Onyenekwe et al. (2000)
Stachyose	Soymilk	Fermentation by *Lactobacillus bulgaricus* and *Streptococcus thermophilus* after eight hours at 44°C	31.5% stachy-ose	Omogbai et al. (2005)
Raffinose and stachyose	Yambea seeds	Fermentation by *E. oligosporus*	98%	Azeke et al. (2007)
Raffinose and stachyose	Soybean flour	α-galactosidase from germinated soybean (after six hours at 40°C)	72.3% stachy-ose 89.2% raffinose	Viana et al. (2005)
Stachyose	Legume flours (field pea and pinto bean)	Fermentation using *Lactobacillus fermentum* or *L. plantarum*	27% (pinto bean flour) 43% (field pea flour)	Duszkiewicz-Reinhard et al. 2002

SOURCES OF α AND β-GALACTOSIDASES

Several sources of α-galactosidases have been studied (Shivanna et al., 1989). According to Manzanares et al. (1998), the bacterial enzyme shows some advantages comparing to fungal α-galactosidases, since its higher production. However, fungal galactosidases are obtained easily due its extracellular location and its wide stability profile, being therefore viable for its production in a large scale (Annunziato et al., 1986; Somiari and Balochi, 1992). This enzyme presents different forms, monomeric or tetrameric, and presents thermal stability (Alani et al., 1989; Porter et al., 1992). The great majority of the enzymes are glycoproteins, with glucose and mannose residues in soya enzyme (Porter et al., 1992; Zeilinger et al., 1993). In soybean, the galactose protects partially the catalytic site in tetrameric form against heat denaturation, but this protection is pH-dependent. Studying the enzyme of vegetable source, Mulimani and Ramalingam (1997) performed experiments with soy flour and concluded that galactosidases from guar (*Cyamopsis tetragonolobus*) better reduced oligosaccharides content than microbial enzymes.

In plants, the activity of β-galactosidases has mainly been described in the processes of growth, development, senescence, and fruit ripening, where they are typically thought to act on cell wall carbohydrates (Chantarangsee et al., 2007; Li et al., 2001).

During soybean seed germination, the content of ROs decreased substantially, while the α-galactosidase activity increased. α-galactosidase was partially purified from germinating seeds by partition in an aqueous two-phase system and ion-exchange chromatography. The enzyme preparation presented maximal activities against q-nitrophenyl-α-D-galactopyranoside (qNPGal) at 60°C and a pH of 5.0 and the K_M values for qNPGal, melibiose, and raffinose of the enzyme preparation were 0.33, 0.42, and 6.01 mM, respectively. The enzyme was highly inhibited by SDS, copper, and galactose (Viana et al., 2005) (see Table 6.3).

Table 6.3. Vegetable sources of galactosidases.

Source	Conditions extraction	Activity	Reference
Papaya	Fruits were retreated with liquid nitrogen, ground to fine powder and homogenized in citrate buffer containing NaCl (1 M), disodium EDTA, PVP and β-mercaptoethanol, in a blender. The homogenate was centrifugated and the supernatant recovered for enzyme assay.	Three galactosidases: α1:1.6 nkatal/ml α2:40 nkatal/ml α3:3 nkatal/ml	Soh, Ali and Lazan (2006)
Erythrina indica three	Mature seeds were soaked in distilled water for overnight.The seed coats were peeled off and the softened cotyledons were homogenized in a mixer with saline (0.85% NaCl, 0.02% sodium azide) (500 mL saline/100 g cotyledons). All steps were carried out at 4–7°C.	10.8 U/ml saline extract	Kestwal and Bhide (2007)
Vicia faba seeds	α-galactosidases	16 nkatal/ml	Dey, Naik and Pridham (1982)
Oryza sativa rice	The rice seeds were soaked for up to 10 days at 28°C under sterile conditions on tissue paper moistened with sterile distilled water. The whole rice plants were harvested and kept at 70°C.	β-galactosidases activity not determined	Chantarangsee et al. (2007)

Table 6.3. *(Continued)*

Source	Conditions extraction	Activity	Reference
Sugarbeet root	Roots were hand washed, diced into small pieces, immersed in liquid N_2, lyophilized, ground to a fine powder and stored at -80°C. Dialyzed extracts were assayed for α-galactosidase activity	α-galactosidase	Haagenson et al. (2008)

α-Galactosidases

The enzyme α-D-galactoside galactohydrolase (EC 3.2.1.22), also named α-galactosidase or α-gal, catalyzes the hydrolysis of α-1, 6-galactosidic bonds, releasing α-D-galactose as the enzymatic reaction product. These bonds are found in oligosaccharides such as melibiose, raffinose and stachyose (Fig. 6.1), which are associated to flatulence in monogastric animals and humans (Suarez et al., 1999). Human consumption of soy-derived products has been limited by the presence of these nondigestible oligosaccharides (NDOs), since most mammals, including man, lack pancreatic α-galactosidase. However, such NDOs can be fermented by gas producing microorganisms present in the cecum and large intestine, thus inducing flatulence and other gastrointestinal disorders in sensitive individuals (LeBlanc et al., 2004). α-galactosidases are applied in biocatalytic processes to improve quality to products destined to human nutrition and also as a component in animal diets to increase digestibility and reduce the fermentation of NDOs. Nutritional studies using commercial preparations of α-galactosidase as a supplement in the feed of monogastric animals demonstrated to significantly increase in weight profit simultaneously with digestibility in swine and chickens. The food industry, especially the soy derivatives sector, has great interest in the reduction of galacto-oligosaccharides, which are resistant to heat and, therefore, are not eliminated during conventional processing (Said and Pietro, 2004). Mulimani et al. (1997) reported that crude α-galactosidase treatment on soybean flour reduced the raffinose and stachyose contents by 90.4% and 91.9%, respectively. In the beet-sugar production process, this enzyme hydrolyses raffinose present in the beetroot must be boiled, contributing to product crystallization (see Fig. 6.2).

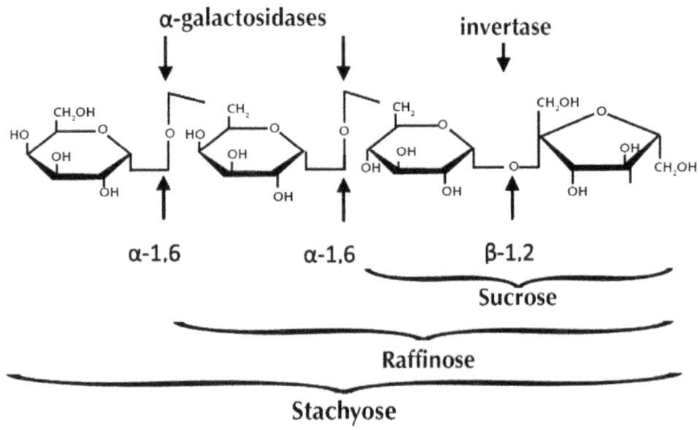

Figure 6.2. Structure of galactosides (raffinose and stachyose) and the enzymes that breaks the glucosidic linkage (based on LeBlanc et al., 2004; Sanada et al., 2009).

α-Galactosidase is abundant in plants, microorganisms, and animals. It hydrolyzes a variety of simple α-D-galactoside, as well as more complex molecules such as oligosaccharides and polysaccharides (Cruz & Park, 1982;). Several groups have developed enzymatic processes using α-galactosidase from plant, bacterial, and fungal source for the hydrolysis of galacto-oligosaccharides of raffinose family.

α-Galactosidases present in bean grains can be potentially used for several biotechnological applications, mainly in the food industry to hydrolyze NDOs in soybean products (Guimarães, 2001; Viana et al., 2005). Fialho et al. (2006) demonstrated the specific α-galactosidase activity in the bean grain. The values varied from 0.032 to 0.049 U/mg and suggested that the varied "Mexico 222" is a promising source of α-galactosidase for industrial purposes.

α-GALACTOSIDASE PRODUCTION BY MICROORGANISMS

α-Galactosidases are widely distributed in microorganisms, plants, and animals (Said and Pietro, 2004). Microorganisms are able to produce α-galactosidases with high yield. The production of these enzymes by microorganisms as fungi, yeasts, and bacteria consists of microorganism growth in a selective and generally liquid culture medium, containing an enzymatic activity inductive carbon source. Sugars as galactose, melibiose, raffinose, and stachyose have already been described as inductors of α-galactosidase activity (Rezende et al., 2005; Said and Pietro, 2004).

Sakai et al. (1987) studied the performance of *Bifidobacterium breve* in soy milk and observed raffinose and stachyose reduction. The maximal enzyme activity was detected at 50°C at pH 5.0. These optimal conditions are similar to those observed to the enzyme produced by *Bifidobacterium longum* and to *Lactobacillus fermentum* (Garro et al., 1994; Garro et al., 1996). However, differences on enzyme behavior are observed in the presence of inhibitors: *p*-chloromercuribenzoato (PCMB) affected the enzymatic activities of *B. breve* and *L. fermentum* suggesting the presence of sulfidrile groups in the catalytic site of enzyme, while α-galactosidase produced by *B. longum* does not suffer with the addition of PCMB, revealing different structures to enzymes produced by distinct microorganisms. The addition of Hg^{2+} ions inhibits the enzyme activity produced by the three microorganisms due to the presence of tiol groups.

Porter et al. (1992) showed the use of α-galactosidase produced by *Bacillus stearothermophilus* presented thermo stability at 65–70°C. This thermal resistance allows the application of this enzyme in food and feed industries, where thermal processing is required. Similar studies were done by Eveleigh (2000) using the bacteria *Thermotoga neopolitana,* for α-galactosidase to industrial application.

Sanada et al. (2009) studied the production of α-galactosidase by submerged fermentation using soybean vinasse as substrate, a residue of the alcoholic fermentation of soybean molasses. Soybean molasses is a by-product of the protein-concentrate soybean meal production. *Lactobacillus agilis* was selected for presenting the highest enzymatic activity, equivalent to 11.07 U/mL after 144 hr of fermentation. The effects of soluble solids concentration in the fermentation medium, C:N relation, and size of inoculum were investigated. Results demonstrated that the medium concentration of 30% soluble solids, with a C:N relation of 9, and size of inoculum of 25% (v/v) were the best conditions for α-galactosidase production (see Table 6.4).

Table 6.4. α-galactosidases produced by microorganisms.

Microorganism	Conditions	Reference
Debaryomyces hansenii UFV-1	0.62 g/l of KH_2PO_4, 2.0 g/l of K_2HPO_4, 1.0 g/l of $(NH_4)_2SO_4$, 0.1 g/L of $MgSO_4.7H_2O$ and 5.0 g/l of yeast extract with galactose as the carbon source. Incubation at 30°C/ 200 rpm for 31 hr. Supernatant separation by centrifugation to evaluate extracellular enzyme and the biomass, the intracellular enzyme.	Viana et al. (2011)
Aspergillus awamori	Solid state fermentation using raw pinto bean flour contained 0.38% raffinose and 3.34% stachyose on a dry weight basis treated with α-galactosidase (60 U/ml). Complete hydrolysis after two hours	Song et al. (2009)
Lactobacillus agilis	Submerged fermentation, medium at 30% (w/v) soluble solids containing soybean vinasse, yeast extract, C:N 9, 144 hr,	Sanada et al. (2009)
Aspergillus oryzae	Solid state fermentation in tray, red gram plant waste + wheat bran, 81% moisture content, pH 5.5, 4 days; 30°C; 4.37 U/g	Shankar and Mulimani (2007)
Rhizopus sp.	Enzyme-induced medium containing 3% soybean meal, bran, lactose, galactose, xylose, glucose, cellulose. Submerged fermentation at 250 rpm, 30°C, 120 hr, 0.4 U/ml, 78 kDa	Cao et al. (2007)
Bacillus stearothermophilus	Submerged fermentation, 0.2% (w/v) galactose, phosphate salts and yeast extract, pH=7.0, 200 rpm, 60°C, 24 hr	Gote et al. (2006)
Penicillium sp.	Submerged fermentation, 250 rpm, 30°C, 7 days	Mi et al. (2007)
Bifidobacterium lactis	Submerged fermentation in sterilized commercial soymilk for 48 hr at 42°C reduced raffinose in 77.4%	Donkor et al. (2007)
Humicola sp.	Solid state fermentation using soya flour at 45°C, 10 days, initial moisture of 95%, final pH 8.5, 44.6 U/g	Kotwal et al. (1997)
Lactobacillus reuteri	Raffinose as substrate, highest yield at the beginning of the stationary growth phase; 64 kDa enzyme	Tzortzis (2003)
Lactobacillus fermentum Lactobacillus plantarum	Solid state fermentation using pinto bean and fiend pea flours, moisture adjusted to 20%, 72 hr, stachyose reduction of 27%	Duszkiewicz-Reinhard et al. (2006)
Lactobacillus fermentum Lactobacillus plantarum	Solid state fermentation using field pea flours, moisture adjusted to 20%, 72 hr, stachyose reduction of 43%	Duszkiewicz-Reinhard et al. (2006)
Gibberella fujikuroi	Submerged fermentation using Czapek-Dox media with 20 g/l of galactose as carbon source, pH=5.5, 160 rpm, 27°C	Thippeswamy and Mulimani (2002)

Table 6.4. *(Continued)*

Microorganism	Conditions	Reference
Streptomyces griseoloalbus	Solid state fermentation composed by soybean flour supplemented with 1.5% galactose, salts solution, incubated at 30°C, 40% initial moisture, inoculum 1.9×10^6 CFU/g initial dry substrate, 5 days, 117 U g^{-1} of dry fermented substrate.	Anisha and Prema (2008)
Streptomyces griseoloalbus	Submerged fermentation, 20.4 U/ml, initial pH 6.0, 1% locust bean gum, 30°C, 10% (v/v) inoculum	Anisha and Prema (2007)
Mortierella vinacea	α-galactosidases Protein precipitation from mycelium extract with $(NH_4)_2SO_4$ (60-100% of saturation) (0.85 U/g)	Miszkiewicz and Galas (2000)
Pichia guilliermondii	Submerged fermentation; maximal α-galactosidase detected in media containing lactose and the soytone product at pH 7.	Church and Meyers (1980)
Thermomyces lanuginosus	Submerged fermentation 3% of sucrose, 0.9% ammonium acetate, 9 days of fermentation, 90–100 U/ml	Rezessy-Szabó et al. (2003)

Crude cell-free extracts from *Lactobacillus reuteri* grown on cellobiose, maltose, lactose, and raffinose were assayed for glycosidic activities. When raffinose was used as the carbon source, α-galactosidase showed the highest yield at the beginning of the stationary growth phase. A 64 kDa enzyme was purified by ultra- and gel-filtration, and characterized for its hydrolytic and synthetic activity. Highest hydrolytic activity was found at pH 5.0 at 50 C (K_M 0.55 mM, V_{max} 0.80 μmol min^{-1} mg^{-1} of protein). At a substrate concentration of 23% (w/v) oligosaccharide mixtures were formed with main products being a trisaccharide at 26% (w/w) yield from melibiose after eight hours and a tetrasaccharide at 18% (w/w) yield from raffinose after seven hours. Methylation analysis revealed the trisaccharide to be 6′ α-galactosyl melibiose and the tetrasaccharide to be stachyose. In both cases synthesis ceased when hydrolysis of the substrate reached 50% (Tzortzis et al., 2003).

Lactic acid bacteria such as *Lactobacillus plantarum*, *Lactobacillus fermentum*, *Lactobacillus brevis*, *Lactobacillus buchneri,* and *Lactobacillus reuteri* are also able to hydrolyse α-galactooligosaccharides into digestible carbohydrates during vegetable fermentations. Recently the characterization of genes involved in α-galactooligosaccharides hydrolysis by *Lactococcus raffinolactis* was described (LeBlanc et al., 2005). Bacterial enzyme-producing species also include *L. fermentum* (Garro et al., 1996), *Bacillus stearothermophilus* (Gote et al., 2004). Fungal enzymes can be produced by *Aspergillus fumigatus* (Rezende et al., 2005), *Humicola* sp. (Kotwal et al., 1998) and *Aspergillus niger* (Manzanares et al., 1998).

β-GALACTOSIDASES AND THE LACTOSE INTOLERANCE

A great part of world population presents some food intolerance. Food intolerance is an adverse food reaction resulting from enzymatic defects of the gastrointestinal tract.

The symptoms are manifested in the gastrointestinal tract after ingestion of certain foods.

The most common food intolerance is caused by the consumption of lactose. This comes from a genetically predetermined reduction in β-galactosidase activity, a hydrolase also known as lactase which hydrolyzes the carbohydrate lactose from dairy products in the small intestine (Moriwaki and Matioli, 2000; Ortolani and Pastorello, 2006). During the neonatal period, the enzyme is essential in human nutrition. The decline of the intestinal β-galactosidase (β-gal or commonly known as lactase) activity is a biological characteristic of the maturing intestine in the majority of the world's population. This declination, termed hypolactasia, causes insufficient lactose digestion in the small intestine. Hypolactasia and lactose bad absorption accompanied with clinical symptoms, such as bloating, flatulence, nausea, abdominal pain and diarrhoea, are termed lactose intolerance. Numerous studies have shown that individuals with hypolactasia could tolerate fermented dairy products better than an equivalent quantity in milk (Vasiljevic, 2009).

Later, the specific activity of lactase decreases to low adult levels (Moriwaki and Matioli, 2000). It can affect 6–12% of caucasians, but in some ethnic groups the prevalence can exceed 60%. Due to β-galactosidase deficiency lactose cannot be completely hydrolysed and the molecules reach the colon where they are degraded by bacteria in H_2O, CO_2, and H_2.

The lactose that is not easily absorbed in the small intestine passes on to the colon where it produces fluid accumulation by osmotic action and undergoes microbial fermentation causing intestinal distress (Potter and Hotchkiss, 1995). The fermentation in the colon causes disturbances such as bloating, abdominal pain and sometimes diarrhoea (Ortolani and Pastorello 2006). This frequent gastrointestinal symptoms (are abdominal pain, flatulence, or diarrhoea), may result from the fermentation of undigested lactose by colonic bacteria. The reduction or elimination of milk and dairy products from the diet of lactose-intolerant children may compromise their uptaken of protein, riboflavin and calcium. Accordingly, alternatives to milk elimination from the diet have been developed. Microbially produced β-galactosidase, is added to milk to prehydrolyse lactose, or can be consumed when foods containing lactose are ingested. Powdered fermented milk that retains β-galactosidase activity also reduced lactose maldigestion in children. It has been proposed that improved fermented milk products could be produced using strains of bacteria that possess high β-galactosidases activity at acidic pH. Another approach to the management of lactose digestion is the consumption of fermented (unfermented) milk containing live lactic bacteria like *Lactobacillus acidophilus* and *Bifidobacterium longum* (Moriwaki and Matioli, 2000).

It is now well established that population who are lactase non-persistent digest lactose more efficiently from yogurt and fermented milks than from milk. An *in vitro* assay was used to study the lactase activity of *Bifidobacterium bifidum*. Frozen cells and fermented milk supplemented with such living organisms were tested. Lactase activity was measured by lactose disappearance. Glucose and galactose by-products from lactose hydrolysis were used differently by the bacterial cells (Passerat and Desmaison, 1995).

β-galactosidases catalyze the hydrolysis of β(1–3) and β(1–4) galactosyl bonds in oligosaccharides. Also catalyse the inverse reaction of enzymatic condensation and transglycosylation. β-galactosidase facilitates the hydrolysis of lactose in the two monosaccharides (glucose and galactose) which are then absorbed.

Rojas et al. (2004) reported the crystallographic structure of *Penicillium* sp. β-galactosidase, which is a glycoprotein containing seven N-linked oligosaccharide chains.

The lactase hydrolyses the lactose in glucose and galactose, whose edulcorant capacity is 0.8 times of sacarose; the hydrolysed is also 3 to 4 times more soluble than lactose. Besides, monosaccharides are easily absorbed in the intestine (Wiseman, 1991).

β-Galactosidases Production by Microorganisms

Commercial sources of β-galactosidases are *Saccharomyces lactis*, *Aspergillus niger*, *Aspergillus oryzae*, *Escherichia coli*, *Saccharomyces*, and *Kluyveromyces*. The lactase from milk yeast *S. lactis* is the most used commercial enzyme. The optimal activity conditions are 35–40°C and pH 6.6–6.8, are similar to the milk yeasts, but it is mostly used in the milk and non-acid whey treatments. The second known lactase preparation is derived from *A. niger*, whose optimum activity conditions are 50°C and pH 3.5–4.5 to application in acid-whey. The enzymes from *A. niger* and *S. lactis* are inhibited by the product, the galactose, and it is too difficult to get a complete lactose hydrolysis (Wiseman, 1991).

β-Galactosidase from *Kluyveromyces marxianus* was produced under submerged fermentation using a culture medium based on cheese whey. The fermentations were carried out in a shaker at 30°C and initial pH 5.5 under agitation, starting with an initial cellular concentration of 10^7 cells per ml, where the initial concentrations of lactose and yeast extract were varied. Enzymes were extracted from the cells by lysis with chloroform in potassium phosphate buffer. In the medium with an initial lactose concentration of 50 g/l, supplemented with salts, yeast extract 12 g/l, the enzymatic activity, and cellular concentration were 28 U/ml and 5.3g/l respectively (Santiago et al., 2004) (see Table 6.5).

Table 6.5. Microbial sources of β-galactosidases.

Microorganism	Conditions	Enzymatic Activity	Reference
	α-galactosidases		
Mortierella vinacea	Protein precipitation from mycelium extract with $(NH_4)_2SO_4$ (60–100% of saturation)	0.85 U/g	Miszkiewicz and Galas (2000)
Alicyclobacillus acidocaldarius	Submerged fermentation, 100 rpm, 65°C, 24 hr	0.69 U/mg	Gul-Guven et al. (2007)
Bacillus subtilis	Medium containing oat, chestnut flour, tryptona yeast extract	Specific activity: 73 U/mg (oat) 56 U/mg (chestnut flour) 910 U/mg (tryptona)	Konsoula et al. (2007)

Table 6.5. *(Continued)*

Microorganism	Conditions	Enzymatic Activity	Reference
Kluyveromyces marxianus	Submerged fermentation, whey medium with 50 g/l lactose, 12 g/l yeast extract, $(NH4)_2SO_4$, KH_2PO_4 e $MgSO_4.7H_2O$, 150rpm, 30°C, pH 5.5	28 U/ml	Santiago et al. (2004)
Bifidobacterium bifidum	Submerged fermentation using milk, initial pH 6.5. In 240 hr, utilized 5.94 mmol/ 100 ml of lactose, final pH 6.13	—	Passerat and Desmaison (1995)
K. lactis	Submerged fermentation 200 rpm, 30°C, 48 hr, containing (g l^{-1}): lactose, 30; yeast extract, 1; K_2HPO_4, 2; $NH_4H_2PO_4$, 1; $(NH_4)_2HPO_4$, 1; $MgSO_4$. $7H_2O$, 0.1 and pH was adjusted to 7.0, intracellular enzyme	3416 U/mg (specific activity)	Dagbagli and Goksungur (2008)
Bififobacteria longum	Submerged fermentation, 4 % lactose, 3.5 % yeast extract, 0.3% K_2HPO_4, 0.1 KH_2PO_4, 0.05 $MgSO_4.7H_2O$; 0.03% L-systeine, 16 hr of fermentation, 37°C.	18.6 U/mL	Hsu et al. (2005)
Bifidobacteria infantis	Submerged fermentation, 16h of fermentation, pH 6.5, 37°C in a medium containing 4% lactose, 3.5% yeast extract, 0.3% K_2HPO_4 , 0.1% K_2HPO_4 , 0.05% $MgSO_4$ H_2O and 0.03% L-cysteine.	13 U/ml	Hsu et al. (2005)
Kluyveromyces lactis	Solid state fermentation, corn grits and wheat bran, 250 rpm, 30°C, 20 hr	~ 30 U/g	Becerra and Siso (1996)

α-GALACTOSIDASE IN FOODS

α-Galactosidase has been used in some countries for the hydrolysis of raffinose into sucrose and galactose in concentrated sugar syrups and molasses from sugar beet processing. This process provides for improved sugar crystallization and a higher sucrose yield. For example, Hokkaido Sugar Company has treated 300 tons of molasses with 1–5% raffinose per day in 1990 (Buchholz et al., 2005) (see Table 6.6).

Table 6.6. Application of α-galactosidases in food products.

Enzyme	Application	Reference
α-galactosidases from Bifidobacterium	Reduce raffinose and stachyose in soy-based products	Scalabrini et al. (1998) Tsangalis and Shah (2004) Liu (1997)
α-galactosidases from *Aspergillus oryzae* or *Aspergillus soyae*	Reduction of raffinose and stachyose during fermentation of soy sauce	Caplicec and Fitgerald (1999)
α-galactosidases from *R. oligosporus* α-galactosidases from germinated soybeans	Tempeh (an Indonesian soy-based food obtained by solid state fermentation)	Caplice and Fitgerald (1999)
α-galactosidases	Tofu	Shurtleff and Aoyagi (1999)
α-galactosidases from *R. oligosporus*	Yambean based foods (legume fermented)	Azeke et al. (2007)
Lactobacillus fermentum or *Lactobacillus plantarum*	Legume flours (field pea and pinto bean)	Duszkiewicz-Reinhard et al. (2002)
α-galactosidases	Beans	Yamaguishi et al. (2009)

β-GALACTOSIDASES IN FOOD

Relief of the symptoms of lactose intolerance is probably the most widely accepted health benefit of probiotic organisms. The traditional cultures used in making yoghurt (i.e., *L. delbrueckii* sp. *bulgaricus* and Str. *thermophilus*) contain substantial quantities of β-D-galactosidase (Shah, 2007), and so both yoghurt and probiotic yoghurt are tolerated well by lactose intolerant organism. However, reduced levels of lactose in fermented products due to partial hydrolysis of lactose during fermentation are only partly responsible for this tolerance to yoghurt (Bernardeau et al., 2008; Donkor, 2007; Rodgers, 2008).

Foods that contain pro and prebiotics are called symbiotic. An example is the water soluble extract of soybeans containing the oligosaccharides raffinose and stachyose, as prebiotics to stimulate the growth of probiotics, and an alternative used by people who have allergies and/or disturbances caused by the ingestion of milk (Spier et al., 2009).

The high lactose content in milk products such as ice-cream, frozen milks, whey spreads, and condensed milk can lead to excessive lactose crystallization resulting in products with a mealy, sandy, or gritty texture. The application of β-galactosidase to dairy products could reduce lactose concentration to acceptable values, and so improve some technological and sensorial quality of dairy foods; for example increasing the digestibility, softness, creaminess, and so forth (Zadow, 1993).

In the cheese industry whey is a waste containing lactose, which causes many environmental problems. Whey containing high level of lactose tends to crystallize easily. Besides, lactose is a relatively insoluble sugar which provokes several problems during storage and transportation by its crystallization (Wiseman, 1991).

Lactose may be hydrolysed by β-galactosidases into very useful sweet syrup, which can be used in the dairy, confectionary, baking, and soft drinks industries (Pivarnik et al., 1995) (see Table 6.7).

Table 6.7. Application of galactosidases in food products.

Enzyme	Application	Reference
β-galactosidases	Sweet syrup production	Pivarnik et al. (1995)
β-galactosidases	Whey, condensed milk, frozen milk, ice cream	Wiseman (1991)
β-galactosidases produced during fermentation by lactic acid bacteria or probiotics	Dairy products	Shah (2007)
β-galactosidases produced during industrial lactic fermentation of milk	Yoghurt production	Rodgers (2008); Bernardeau et al. (2008); Donkor (2007)
β-galactosidases from *Sulfolobus solfataricus*	Lactulose production	Yeong-Su Kim et al. (2006)
immobilized thermostable β-glucosidases from *Pyrococcus furiosus*	Lactulose production	Mayer J. et al. (2010)
β-galactosidases from *Lactobacillus delbrueckii* subsp *bulgaricus*	Lactose bioconversion	Rhimi M. et al. (2009)
Mesophilic β-galactosidases	Lactose-reduced skim milk production	Novalin S. et al. (2005)

TRENDS ON GALACTOSIDASES APPLICATION

Food industries focus on the production of new products by the evolution of technological processes for extending the shelf-life and increasing human health. Besides, the reduction of collateral and adverse effects to their digestive systems are also based on population demand and necessity. Some components in vegetables such as legumes and grains bring anti-nutritional or indigestible effects to human consumption.

In the last decades, bean processing was basically made by grain processing companies which used basic unit operations such as washing, drying, selection, classification, cooking, and packing to commercialization.

Another way of beans, grains, and legumes industrialization is pre-cooking the grains, drying in a drum dryer, getting flakes to instantaneous preparation. This product is used in soups and composition of culinary preparation. The demand increase in bean derived foods, the industries should apply enzymes that degraded oligosaccharides non-digestible to minimize gastrointestinal problems described above. After some pre-treatment of these vegetables, α-galactosidases are the main group of enzymes that hydrolysis these components in digestible molecules that may permit the use of bean product in baby food after research studies about digestibility of the components in the hydrolyzed product.

The increasing demand by products based in grains such as soy and black beans turns an alternative to added value to these raw materials, promote economy growth, employee generation, and offer to the consumer new alternatives of consumption of healthy foods.

New alternatives for children products and baby foods could be developed using legumes and grains not recommended for children before a year old due to the presence of indigestible compounds. So, vegetables submitted to enzymatic treatment in order to eliminate these compounds such as non-digestible oligosaccharides could be an alternative to the children's food industries and new perspectives and consumption options to this important and promissory sector.

Dairy products with low lactose content are been developed to lactose-intolerant population. Products are obtained by lactic acid bacteria or enzymatic hydrolysis of lactose in milk, where glucose and galactose are release during processing, and such sugars are known as secure and avoid digestion problems related before.

KEYWORDS

- **Galactosidases**
- ***Glycine max***
- **Monosaccharides**
- **Oligossacarides**
- **Tetrassacharide**

Chapter 7

Medicinal Properties of Edible Mushrooms

Carlos Ricardo Soccol, Leifa Fan, and Sascha Habu

INTRODUCTION

Mushrooms belong to Fungi Kingdom although in the past, they were classified as plants. The fruiting bodies receive attention for their forms and colors when they are found in nature. The interest in mushrooms is old, Oriental countries use the mushrooms for the treatment and prevention of several diseases such as arthritis, rheumatism, bronchitis, gastritis, cancer, as well as in health and longevity maintenance. Scientists search answers about the action mechanism of mushrooms and about alternative treatments based on natural compounds. Many species are appreciated in culinary around the world because of their good taste and their nutritious potential. There are researches about bioactive compounds such as polysaccharides, proteo-glucans, phenol compounds, nucleotides and their action mechanism to treatment for several diseases. Studies have shown significant results in the immune system improvement, antimicrobial and anti-angiogenic activity and in the treatment of the cancer and high cholesterol. The aim of this chapter is describes any biological properties of mushrooms to the treatment of several diseases.

Mushrooms are a special group of macroscopic fungi. Macromycetes arranged in the phylum Basidiomycota and some of them in the Ascomycota are known as the higher fungi (Moradali et al., 2007, Sicoli et al., 2005). It is considered that exist in the planet about 140,000 different species of mushrooms, however, only about 10% are known. Half of them present nutritious properties, 2000 species are safe and approximately 70 of them are known by presenting some pharmacological property.

Edible mushrooms are attractive because of their flavor, taste, and delicacy (Diyabalanage et al., 2008). Although many species of edible mushrooms exist in the nature, less than 20 species are used as food and only 8–10 species are regularly cultivated in significant extent.

Fresh mushrooms can be acquired from grocery stores and markets, including straw mushrooms (*Volvariella volvacea*), oyster mushrooms (*Pleurotus ostreatus*), shiitakes (*Lentinula edodes*) and enokitake (*Flammulina* sp.). There are many other fungi like milk mushrooms, morels, chanterelles, truffles, black trumpets and porcini mushrooms (*Boletus edulis*, also known as "king boletes") (Ghorai et al., 2009).

Many worldwide cultures, especially in the Orient, recognize that extracts from some edible and non-edible mushrooms were recognized for their potential health benefits. In China, the dietary supplements and nutraceuticals made from mushroom extracts are used, along with various combinations of other herbal preparations (Barros et al., 2008a; Carbonero et al., 2006).

These mushrooms have attracted attention because they are source of non-starchy carbohydrates, with a high content of dietary fiber (chitimous wall), moderate quantities of proteins (20–30% of dry matter) with most of the essential amino acids, minerals and vitamins (B) (Agrahar-Murugkar and Subbulakshmi, 2005; Ghorai et al., 2009).

Various compounds with important pharmaceutical properties have been isolated from these organisms. Substances that act as anti-aging, in longevity, modulating the immune system, having hypoglycemic activity and to inhibit tumor growth have been isolated from mushrooms, such as polysaccharides. Polysaccharides can interconnect several points forming a wide variety of branched or linear structures, for example, ß-glucans. Polysaccharides have structural variability for regulatory mechanisms of various interactions in higher organisms (Agrahar-Murugkar and Subbulakshmi, 2005; Carbonero et al., 2006). Furthermore, other bioactive substances such as triterpenes, lipids and phenols have also been identified and characterized in mushrooms with medicinal properties (Maiti et al., 2008).

Fungi can be produced technically through fermentative process. The media may be in form of available substrates from valued cheap sources like agro-biomass and industrial waste; transformed into high value added food and pharmaceutical products.

BIOLOGICAL PROPERTIES OF MUSHROOMS

Antioxidants Activity

Exogenous chemical and endogenous metabolic process in the human body or in the food system might produce highly reactive free radicals, especially oxygen derived radicals, and especially oxygen derived radicals, which are capable of oxidizing biomolecules, resulting in cell death and tissue damage (Elmastas et al., 2007). During the reduction of molecular oxygen, reactive oxygen species are formed and there is a continuous requirement for inactivation of these free radicals. Superoxide and hydroxyl radicals are the two most representative free radicals. In cellular oxidation reactions, superoxide radical is normally formed first, and its effects can be magnified because it produces other kinds of cell-damaging free radicals and oxidizing agents. Damage induced by free radicals can affect many biological molecules, including lipids, proteins, carbohydrates and vitamins present in food. Reactive oxygen species also implicate in the pathogenesis of various human diseases, such as DNA damage, carcinogenesis, rheumatoid arthritis, cirrhosis, arteriosclerosis as well as in degenerative processes associated with ageing. Evidences have been indicating that diet rich in antioxidant reduce risks of some diseases (Elmastas et al., 2007; Liu et al., 1997). Mushroom contain vitamins A and C of ß-carotene and a great variety of secondary metabolites such as phenolics compounds, polyketides, terpenes, steroids and phenols, all have protective effects because of their antioxidant properties (Jayakumar et al., 2009; Soares et al., 2009).

Researchers investigate several edible and non-edible mushrooms with antioxidant properties for applications in food, cosmetics and treatment of diseases (Table 7.1). Water or ethanolic extracts of fruiting bodies or biomass resulting by fermentation have been studied and tested.

Table 7.1. Potential antioxidant of mushrooms.

Mushroom	Biological Activity: Antioxidant		
	Source	**Substance**	**References**
Agaricus bisporus	Fruit body	Methanolics	Elmastas et al. (2007)
Agaricus brasiliensis	Fruit body	Methanolic	Soares et al. (2009)
Agaricus silvaticus	Fruit body	Methanolic	Barros et al. (2008b)
Agrocybe cylindracea	Fruit body	Ethanolic and hot water extracts	Tsai et al. (2007)
Antrodia camphorata	Submerged Fermentation	Methanolic	Shu and Lung (2008)
Boletus edulis	Fruit body	Hot water extracts	Ribeiro et al. (2008)
Boletus edulis	Fruit body	Alkaloids	Sarikurkcu et al. (2008)
Boletus badius	Fruit body	Methanolic extracts	Elmastas et al. (2007)
Cordyceps sinensis	Submerged Fermentation	Polysaccharide	Leung et al. (2009)
Geastrum saccatum	Fruit body	Glucans	Dore et al. (2007)
Grifola frondosa	Fruit body	Water extracts	Lee et al. (2008b)
Hypsizigus marmoreus	Fruit body	Cold and Hot water, Ethanolic	Lee et al. (2007)
Inonotus obliquus	Fruit body	Methanolic	Lee et al. (2007b)
Laetiporus sulphureus	Fruit body	Ethanolic	Turkoglu et al. (2007)
Lentinula edodes	Fruit body	Ethanolic	Zheng et al. (2005)
Leucopaxillus giganteus	Fruit body	Methanol	Barros et al. (2008b)
Lepista nuda	Fruit body	Methanolic	Elmastas et al. (2007)
Phellinus linteus	Fruit body	Ethanolic	Song et al. (2003)
Pleurotus ostreatus	Fruit body	Methanolic	Elmastas et al. (2007)
Pleurotus ostreatus	Fruit body	Ethanolic	Jayakumar et al. (2009)
Polyporus squamosus	Fruit body	Methanolic Extracts	Elmastas et al. (2007)
Russula delica	Fruit body	Methanolic	Elmastas et al. (2007)
Suillus collitinus	Fruit body	Methanol extracts	Sarikurkcu Tepe and Yamac (2008)
Turbinaria conoids	Fruit body	Fucoidan	Chattopadhyay et al. (2009)
Verpa conica	Fruit body	Methanolic extracts	Elmastas et al. (2007)

The Authors

Mushrooms are currently available in Taiwan, including *Agaricus blazei*, *Agrocybe cylindracea* and *Boletus edulis*. Ethanolic extracts were more effective than hot water in antioxidant activity using the conjugated diene method and scavenging ability on 2,2-diphenyl-1-picrylhydrazyl (DPPH) radicals whereas hot water extracts were more effective in reducing power, scavenging ability on hydroxyl radicals, and chelating ability on ferrous ions (Tsai et al., 2008). According to Sarikurkcu et al. (2008) studies about the antioxidant activity of *Lacttarius deterrimus*, *Suillus collitinus*, *Boletus edulis*, *Xerocomus chrysenteron*. *L. deterrimus*. *B. edulis* showed activities as strong as the positive controls. The reducing power of the species was excellent. Chelating capacity of the extracts was proportional to the increasing concentration.

Northeast of Portugal is recognized as one of the richest regions of Europe in wild edible mushroom species, which have considerable gastronomic relevance. *Russula cyanoxantha*, *Amanita rubescens*, *Suillus granulates* and *Boletus edulis* are among more common and marketed species. Four species studied are rich in organic acids, but phenolics compounds are present in low amounts. Organic acids, phenolics compounds and alkaloids composition is insufficient to justify the antioxidant potential of analyzed species. Other compounds also participate in the observed activity. These species present antioxidant potential, especially high for *Boletus edulis*, which is also the richest specie in alkaloids (Ribeiro et al., 2008). *Boletus edulis* was a popular edible mushroom in Europe, North America and Asia. This mushroom presented interesting results for antioxidant effects cited by three authors from different regions of world.

Metabolisms, physiology of mushroom and environmental conditions are important and determinant to the production of antioxidant compounds. However, the extraction methodologies are primordial to remove compounds produced intracellular and/or others substances resulting of metabolism (Table 7.2).

Table 7.2. Contents of total phenols in extracts from mushrooms.

Mushroom	Extraction (mg/g)			References
	Hot Water	Ethanolic	Methanolic	
Agaricus arvensis	-	-	2,72	Barros et al. (2008b)
Agaricus bisporus	-	-	4,49	Barros et al. (2008b)
Agaricus romagnesii	-	-	6,18	Barros et al. (2008b)
Agaricus silvaticus	-	-	8,95	Barros et al. (2008b)
Agaricus silvicola	-	-	6,45	Barros et al. (2008b)
Agaricus blazei	5,67	5,80	-	Tsai et al. (2007)
Agrocybe cylindracea	5,6	5,7	-	Tsai et al. (2007)
Boletus edulis	5,81	5,73	-	Tsai et al. (2007)
Hypsizigus marmoreus	10,01	6,89	-	Lee et al. (2008)
Clitocybe maxima (cap)	9,71	9,66	-	Tsai et al. (2008)
Clitocybe maxima (stipe)	5,1	5,51	-	Tsai et al. (2008)
Pleurotus ferulae	7,73	6,71	-	Tsai et al. (2008)
Pleurotus ostreatus	11,1	7,11	-	Tsai et al. (2008)
Pleurotus ostreatus	-	5,49		Jayakumar et al. (2009)

The Authors

The antioxidant activity of five mushroom species: *Agaricus bisporus*, *Agaricus arvensis* Schaeffer, *Agaricus romagnesi* Wasser, *Agaricus silvaticus* Schaeffer and *Agaricus silvicola* were analyzed. All the species proved to have antioxidant activity, especially *A. silvaticus* (Barros et al., 2008b).

Soares et al. (2009) investigated the young and mature extracts of fruiting bodies of *Agaricus brasiliensis.* Both extracts showed antioxidants activities, except the chelating ability ferrous ions. Consumption of fruiting bodies of *A. brasiliensis* might be beneficial to the human antioxidant protection system against oxidative damage.

Hypsizigus marmoreus (peck) Bigelow, also known bunashimeji and hon-shimeji is a mushroom cultivated and commercially available in Taiwan. Naturally occurring antioxidants components, including tocopherols and total phenols, were found in extracts from *H. mamoreus*. The major antioxidant components found in hot water extracts were total phenols and in ethanolic extracts were total tocopherols (Lee et al., 2008a).

Extraction use polar and non-polar solvents and different temperatures, considering solubility and thermostability of each substance, respectively (Table 7.2).

Shiitake (*Lentinus edodes*) have several compounds including bioactive polysaccharides (lentinan), dietary fiber, ergosterol, vitamin B_1, B_2 and C and minerals have been isolated from de fruiting body, mycelia and culture medium of this mushroom. Heat treatment of Shiitake sample increased the overall content of free polyphenolic and flavonoid compounds. The heat treatment can produce changes in their extractability due to disruption of the cell wall thus bound polyphenolic and flavonoid compounds may be released more easily relative to those of raw materials (Choi et al., 2006).

Clitocybe maxima, *Pleurotus ferulae* and *Pleurotus ostreatus* were used to study antioxidant properties. Ethanolic extracts and hot water extracts from *P. ferulae* and *P. ostreatus* were more effective than *C. maxima* cap and stipe antioxidants activities. Total phenols were major naturally occurring antioxidant components found in the range of 5.51–9.66 gallic acid equivalents/g and 5.10–11.1 mg gallic acid equivalents/g for ethanolic and hot water extracts, respectively. These mushrooms could be used in grams levels as food or food ingredient. Therefore, these three mushrooms might serve as possible protective agents in human diets to help human reduce oxidative damage (Tsai et al., 2008).

The ethanolic extract of the *Pleurotus ostreatus* and various known antioxidants showed concentration-dependent anti-oxidant activity by inhibiting lipid peroxidation, scavenging hydroxyl and superoxide radicals, reducing power and chelating ferrous ions when compared to different standards such as ascorbic acid, BHT and EDTA (Jayakumar et al., 2009).

Anti-inflammatory Activity

Inflammatory response is succession of cellular reactions involving the generation and release of cellular mediators such as cytokines. The excessive amount or duration of the production of cytokines, especially TNF-α, can cause serious harm to the body (Dudhgaonkar et al., 2009).

Excessive or unregulated production of these mediators has been implicated in mediating a number of diseases including rheumatoid, arthritis, osteoarthritis, sepsis, chronic pulmonary inflammatory disease, Crohn's disease, ulcerative colitis, and also carcinogenesis. Inflammation is inherent to pathogenesis of a variety of diseases.

Inhibition of activation and the proliferation of these inflammatory cells appears to be an important therapeutic target for small molecular drugs in the treatment of inflammatory diseases and cancer (Dudhgaonkar et al., 2009; Van et al., 2009).

TNF-α is a major pro-inflammatory cytokines with diverse biological activities. Large quantities of TNF-α may induce intravascular thrombosis, shock and cachexia.

Macrophages play important role in host-defense mechanism, and inflammation. The overproduction of inflammatory mediators by macrophages has been implicated in several inflammatory diseases and cancer. The activation of macrophages is important in the instigation of defensive response such as the production of interleukins IL-1ß, IL-6, TNF-α, reactive oxygen species, prostaglandin and nitric oxide. IL-1 ß are also a multifunctional cytokine which has been implicated in pain, fever, inflammation and autoimmune conditions. It stimulates acute phase protein synthesis in the liver and may cause rise in body temperature. It is also up-regulated in many inflammatory diseases. IL-6 is a multifunctional cytokine with pro-/anti-inflammatory properties.

Nitric oxide is an important messenger in diverse pathological functions, including neuronal transmission, vascular relaxation, immune modulation and cytotoxicity against tumor cells. Nitric oxide, secretor product of mammalian cells, produced by inducible nitric oxide synthase, endothelial nitric oxide synthase, and neuronal nitric oxide synthase is considered an important signaling molecule in inflammation (Van et al., 2009). Mushrooms have been applied in the treatment of infections in popular culture or medicine in many countries, such as China, Japan, Russia and Brazil. The anti-inflammatory properties of mushrooms have interested researchers and motivated the investigation of some species (Table 7.3).

Table 7.3. Potential anti-inflammatory of mushrooms.

Mushroom	Anti-inflammatory Activity		
	Source	**Substance**	**References**
Agrocybe cylindracea	Fruit body	Agrocybin	Ngai et al. (2005)
Amanita muscaria	Fruit body	Hot water, methanolic and ethanolic extracts	Michelot and Melendez-Howell (2003)
Fomitopsis pinicola	Submerged fermentation	Polysaccharides	Cheng et al. (2008)
Ganoderma lucidum	Fruit body	Triterpene	Dudhgaonkar et al. (2009)
Geastrum saccatum	Fruit body	Glucans	Guerra Dore et al. (2007)
Inonotus obliquus	Fruit body	Hot water	Van et al. (2009)
Phellinus linteus	Fruit body	Butanol fraction	Kim et al. (2004)
Poria cocos	Submerged fermentation	Polysaccharides	Lu et al. (2009)
Pleurotus nebrodensis	Fruit body	Nebrodolysin	Lv et al. (2009)
Pleurotus pulmonarius	Fruit body	Polysaccharides	Smirdele et al. (2008)

The Authors

The *Geastrum saccatum*, a mushroom native from Brazil, is produced under natural conditions in the unexplored reserve of "Mata da Estrela-Rio Grande do Norte." This basiodiomycete is a saprobic fungus and it is well adapted to tropical regions. The mushroom, known as "Star of the Land," is used in popular medicine by obstetricians and healers, and has curative properties for eye infections and diseases, such as asthma. The anti-inflammatory effects of glucans *G. saccatum* extract on carragennan-induced pleurisy and observed that the glucans extract decreased the number of cells from pleural fluid rats. There is evidence that inhibition of nitric oxide synthase reduces the production of prostaglandins by ciclooxigenase through reduced synthesis of oxide nitric, these decrease several inflammatory symptoms such as vessel dilation. Thus, it could be related to inhibition of diapedesis of cells as mononuclear leukocytes in the inflammation site when used the glucans extract. The animals treated with glucans extract decreased oxide nitric. These effects suggest an anti-inflammatory effect of glucans *G. saccatum* extract (Dore et al., 2007).

Lu et al. (2009) demonstrated that *Poria cocos*, called Fu Ling in China, can participate in the regulation of the anti-inflammatory process. *Poria cocos* is commercially available and is popularly used in the formulation of nutraceuticals, tea supplements, cosmetics, and functional foods in Asia. Chemical compounds found in *Poria cocos* include triterpenes and ß-pachyman, a polysaccharide composed of ß-pachimarose, pachymic acid, and poricoic acid. IFN-v is one of the major mediators which predispose endothelial cells toward inflammatory/immunological responses. The pretreatment with the polysaccharide extracted of *Poria cocos* was dose-dependently and inhibited IFN-v-induced inflammatory gene IP-10 protein release. It suggests that the effect of polysaccharide on IP-10 expression was regulated at the translational level and thus it may participate in regulating inflammatory-related diseases. This polysaccharide showed no toxicity to endothelials cells, indicating the safety of its use.

Inonotus obliquus also known as Chaga, is a black mushroom that grows on birch trees in northern climates such as in Russia. These mushrooms act as traditional medicine to treat gastrointestinal cancer, cardiovascular disease and diabetes. Polyphenolic compounds produced by Chaga can protect cells against oxidative stress. It also showed to inhibit platelet adhesion and aggregation, which plays an important role in thrombosis. Those platelets are important in hemostasis and modulation of the inflammatory response, including the released of cytokines, Chaga may be involved in various aspects in the inflammatory. Levels of nitrite, which is an indicator of oxide nitric concentration, displayed a significant decline when treated with Chaga. The inflammatory effect caused by Chaga may be a cascade effect with inhibition of oxide nitric production (Van et al., 2009).

Dudhgaonkar and researchers (2009) showed that triterpene extract from medicinal mushroom *Ganoderma lucidum* markedly suppressed in the inflammatory response in LPS-active murine macrophages. Specifically, triterpene by *Ganoderma lucidum* suppressed LPS-dependent secretion of TNF-α, IL-6, oxide nitric and prostaglandin E2 from murine macrophages cells. The inhibition of production of oxide nitric and prostaglandin E2 by *Ganoderma lucidum* triterpene was mediated through the down-regulation of expression of Inductible Nitric Oxide Synthase and ciclooxigenase-2, respectively. Moreover, this triterpene inhibited LPS-dependent induction of NF-κB as well as expression, phosphorylation and nuclear translocation of p65 NF-κB subunit. Also, triterpene of *Ganoderma lucidum* seem to be potent in suppressing the key molecules responsible in the inflammatory response. Extract of *Ganoderma lucidum* containing triterpenes or isolated triterpenes (ganoderic acid A, F, DM, T-Q, lucidenic acid A, D_2, E_2, P, methyllcidenate A, D_2, E_2, Q and 20-hydroxylucidenic acid N) suppressed ear-edema inflammation in laboratories animals.

Phellinus linteus, traditional mushroom medicine in oriental countries, showed topical anti-inflammatory activity. Extract ethanolic was evaluated using croton oil-induced ear edema test and showed an inhibitory effect on inflammation. Among the subfractions, the butanol fraction appeared to be most effective in anti-inflammation, supposing that *Phellinus linteus* have hydrophilic compounds (Kim et al., 2004).

Curiously, *Amanita muscaria* is not considerable edible mushroom, but has been used for various purposes, mostly as a psycostimulant, by different ethnic groups from Mexico to Siberia to Eastern Asia. Slavic nations have their own traditions of *Amanita muscaria* use. Ethnic people are especially fond of the beneficial effects they achieve from topical application of ethanolic extract (or strong vodka). Hot water, methanolic and ethanolic extracts of *Amanita muscaria* have description of their use for reduction of the consequences of inflammatory processes in cases of rheumatic diseases, body injuries, insect bites, others (Michelot and Melendez-Howell, 2003).

Antimicrobial Activity

Antibiotic agents have been effective therapeutic use since their discovery in the 20th century. However, it has paradoxically resulted in the emergence and dissemination of multi-drug and resistant pathogens. Antibiotic resistance represents a prospect of therapeutic failure for life-saving treatments (Hearst et al., 2009).

The search for new drugs that is able to inhibit the antibiotic resistance of bacteria. Mushrooms interestingly showed antimicrobial activities, some examples are Table 7.4. Biologist and others researches related that some mushrooms need special attention in natural environmental because of the relation with other species, growing local, conditions of temperature, substrates and others factors environmental.

Table 7.4. Potential antimicrobial of fruit body of mushrooms.

Biological Activity: Antimicrobial		
Mushroom	**Substance**	**References**
Ganoderma japonicum	Oil essential	Liu et al. (2009)
Ganoderma lucidum	Water extracts	Wu et al. (2006)
Ganoderma lucidum	Ganodermin	Wang and Ng (2006)
Laetiporus sulphureus	Ethanol extracts	Turkoglu et al. (2007)
Lentinula edodes	Water extracts	Hearst et al. (2009)
Lentinula edodes	Chloroform extract	Hirasawa et al. (1999)
Leucopaxillus giganteus	Methanol	Barros et al. (2008b)

Table 7.4. *(Continued)*

Biological Activity: Antimicrobial		
Mushroom	**Substance**	**References**
Pleurotus sajor-caju	Ribonuclease	Ngai and Ng (2004)
Russula delica	Ethanol extracts	Yaltirak et al. (2009)
Russula paludosa	Lacase	Wang et al. (2007)
Tricholoma giganteum	Trichogin	Guo et al. (2005)

The Authors

Laetiporus sulphureus (Bull.) Murrill is a wood-roting basidiomycete, growing on several tree species and producing shelf-shaped fruit-bodies of pink-orange color, except for the fleshy margin which is bright yellow. *Laetiporus* species contain N-methylated tyramine derivatives, polysaccharides, a number of lanostane, triterpenoids, laertiporic acids and other metabolites have reported that laetiporic acids might have potential as food colorants. The antimicrobial effect of ethanol extracts of *L. sulphureus* was tested against six species of Gram-positive bacteria, seven species of Gram-negative bacteria and one species of yeast. The most susceptible bacterium was *Micrococcus flavus* (23 ± 1 mm diameter). The ethanol extract of *L. sulphureus* showed no antibacterial activity against *Klebsiella pneumonie* at the concentration used. The ethanol extract exhibited high anticandidal activity on *C. albicans* (Turkoglu et al., 2007).

Russula delica Fr. is used as food in Turkey and growth under coniferous and deciduous trees. The antimicrobial effect of ethanolic extract of *R. delica* was tested against three species of Gram positive bacteria and six species of Gram negative bacteria. Results showed inhibitory activity against *Shigella sonnei* and *Yersinia enterocolitica.* Natural Antimicrobials agents are more safety to the people and low risk for resistance development by pathogenic microorganisms (Yaltirak et al., 2009).

Hirasawa et al. (1999) studied the antibacterial activity of shiitake extracts (*Lentinula edodes*) as a preventive agent against dental caries and adult periodontitis. Shiitake extracts were antibacterial effective against *Streptococcus* sp., *Actinomyces* sp. *Lactobacillus* sp., *Prevotella* sp., and *Porphymonas* sp., of oral origin. This extract of mushroom can be used to prevent dental caries and periodontitis because also supports the idea that inhibit the formation of water-insoluble glucans from sucrose by glucosyltransferase.

Rao et al. (2009) studied the activities of shiitake freeze-dried powder. Bioassay of the extracts showed that all the fractions exhibited qualitative inhibitory activity against bacteria and fungi. Thirty-four compounds from extracts of the shiitake was identified, for example: Cycloheximide (antibiotic that acts as a plant growth regulator, but causes human liver toxicity and reduction in protein synthesis); Bostrycoidin (bioactive *in vitro* against *Mycobacterium tuberculosis*); Anticarcinogênica alkaloids (muscarine, choline); Tanins (epiafzelechin); Terpenoids (adiantone); Cyclopiazonic acid (a natural food contaminant); Aspergillomarasmine; Disulphides, lenthionine compounds in the organic extracts.

Shiitake mushroom extract had extensive antimicrobial activity against 85% of the organisms it was tested on, including 50% of the yeast and mould species in the trial. This compared with the results from both the positive control (Ciprofloxacin) and Oyster mushroom, in terms of the number species inhibited by the activity of the metabolites inherent to the shiitake mushroom (Hearst et al., 2009).

Mushroom proteins have important play antimicrobial activity. Mushroom compositions have contained 2–40% protein according with species. Each protein of mushroom has specific sequence of amino acids and weight molecular. Antifungal proteins have function of protecting organisms from the deleterious consequences of fungal assault; they display a spectacular diversity of structures. An antifungal peptide with a molecular of 9 KDa was isolated from fresh fruiting bodies of the mushroom *Agrocybe cylindracea.* The antifungal peptide, designated as agrocybin, exhibited remarkable homology to RPI 3, a cysteine-rich-protein that is expressed during fruiting initiation in *Agrocybe chaxingu.* Agrocybin is also similar to grape (*Vitis vinifera*) antifungal peptide in N-terminal sequence. The data suggest that antifungal function of agrocybin is important during fruiting initiation for protecting the fruiting bodies. Agrocybin inhibits mycelial growth in *Mycosphaerella arachidicola*, in line with the majority of fungal proteins and peptides that exert their antifungal action against a number fungal species and is not effective against a variety of bacteria (Ngai et al., 2005).

Pleurotus nebrodensis produce hemolysin that can be implicated as a virulence factor. This hemolysin was named nebrodolysin and showed antiviral activity, inclusive exhibits a suppressive action on HIV-1 and reproducible antiviral effect. The mechanism of the antiviral suggested that nebrodeolysin might act in a different way by interaction the infection of the virus (Lv et al., 2009).

Antifungal protein from *Ganoderma lucidum*, ganodermin, inhibits mycelial growth in the phytopatogenic fungi *Botrytis cinerea*, *Fusarium oxysporum* and *Physalospora piricola* (Wang et al., 2007).

Niohshimeji (*Tricholoma giganteum*), produced trichogin, antifungal proteins monomeric and have N-terminal sequence. This protein showed antifungal activity against *Fusarium oxysporum*, *Mycosphoerella arachidicola* and *Physolospora piricola.* The antifungal activity of protein is high compared with others antifungal proteins. Trichogin inhibits HIV-1 reverse transcriptase (Guo et al., 2005).

Ngai and Ng (2004) demonstrated antimicrobial activity of ribonuclease from the extract of *Pleurotus sajor-caju.* The ribonuclease inhibited mycelial growth in the fungi *Fusarium oxysporum* and *Mycosphaerella arachidicola* and bacteria as *Pseudomonas aeruginosa* and *Staphylococcus aureus.* The molecular mass of *P. sajor-caju* RNase is 12 KDa and poly U-specific and high activity, compared with others mushrooms. The N-terminal sequence of *P. sajor-caju* RNase bears resemblance to the terminal sequence of a bacteriocin peptide and to a portion of the sequence in two enzymes involved in RNA-specific editase. This structural feature of *P. sajor-caju* RNase may be related to its antibacterial and RNase activities.

Tricholoma giganteum produce lacase and characterized with N-terminal sequence that dissimilar from reported N-terminal sequences of mushroom lacases. Its molecular mass (43 kDa) is smaller than most of the reported mushroom lacases which are around 60 kDa (Wang and Ng, 2004)

Russula paludosa is a wild edible mushroom collected from Chine. Its fruiting bodies are abundant in the summer. Extracts from fruiting bodies of *R. paludosa* exhibited an inhibitory effect on HIV-1. The peptide was devoid of hemagglutinating, ribonuclease, antifungal, protease, protease inhibitor and lacase activities (Wang et al., 2007).

Ganoderma japonicum is found in China and has been used for the treatment of various diseases. The essential oil of *G. japonicum* has pharmacologicals effects and contains bactericidal components, such as (E)-nerolidol, linalool and (2E, 4E)-decadienal. The antimicrobial results indicated that oil inhibited mainly Methicillin-resistant *Staphylococcus aureus* (MRSA). This component has been confirmed to have bacteriostatic and bactericidal activity, causing changes in cell membrane permeability and bacterial death (Liu et al., 2009).

Antitumoral Activity

The National Cancer Institute (US National Institutes of Health) define Cancer is a term used for diseases in which abnormal cells divide without control and are able to invade other tissues. Cancer cells can spread to other parts of the body through the blood and lymph systems. There are more than 100 different types of cancer. Cancer types can be grouped in main categories (Table 7.5):

Table 7.5. Classification of cancer types.

Categories	Definition
Carcinoma	Begins in the skin or in tissues that line or cover internal organs
Sarcoma	Begins in bone, cartilage, fat, muscle, blood vessels, or other connective or supportive tissue
Leukemia	Starts in blood-forming tissue such as the bone marrow and causes large numbers of abnormal blood cells to be produced and enter the blood
Lymphoma and myeloma	Begin in the cells of the immune system
Central nervous system cancers	Begin in the tissues of the brain and spinal cord

Source: http://www.cancer.gov – September, 2009.

According to World Health Organization (WHO) related lung, stomach, liver, colon and breast cancer cause the most cancer deaths each year. About 30% of cancer deaths can be prevented. It is estimated that in 2030 there will be 26 million cases of cancer worldwide. Deaths from cancer worldwide are projected to continue rising, with an estimated 12 million deaths in 2030 (http://www.who.int/en/).

International Union Against Cancer reported that Africa is less than 0.1% and Asia is only 8.5% of the population is covered by cancer registration. The Chernobyl disaster was a nuclear reactor accident that occurred on April 26, 1986 at the Chernobyl Nuclear Power Plant in Ukraine and it is now estimated that by 2065 there will be 16,000 cases of thyroid cancer and 28,000 cases of other cancers in Europe as a result of this accident (http://www.uicc.org/). Numerous mushroom species are studied and purified substances such as polysaccharides and proteo-polysaccharides are recognized to be the potent immunomodulatory and antitumor (Table 7.6).

Table 7.6. Potential antitumoral of mushrooms.

Mushroom	Biological Activity: Antitumoral		
	Source	**Substance**	**References**
Agaricus brasiliensis	Submerged fermentation	Polysaccharides	Fan et al. (2007)
Agrocybe aegerita	Fruit body	Methanolic extracts	Diyabalanage et al. (2008)
Albatrellus confluens	-	Grifolin	Ye et al. (2007)
Cordyceps sinensis	Submerged fermentation	Polysaccharide	Yang et al. (2005)
Coriolus versicolor	Fruit body	Terpenoids and polyphenols	Harhaji et al. (2008)
Fomes fomentarius	Fruit body	Polysaccharides	Chen et al. (2008)
Ganoderma capense	Fruit body	Lectin	Ngai and Ng (2004)
Ganoderma lucidum	Solid fermentation	Polysaccharides	Rubel et al. (2008)
Ganoderma tsugae	Submerged fermentation	Polysaccharide	Peng et al. (2005)
Grifola frondosa	Submerged fermentation	Polysaccharides	Cui et al. (2007)
Inonotus obliquus	Submerged fermentation	Polysaccharide	Kim et al. (2006)
Lentinula edodes	Fruit body	Fiber	Choi et al. (2006)
Lentinula edodes	Fruit body	Polysaccharides	Frank et al. (2006)
Lentinula edodes	Fruit body	Ethanolic	Hatvani (2001)
Poria cocos	Submerged fermentation	Polysaccharides	Huang et al. (2007)
Pleurotus citrinopileatus	Fruit body	Lectin	Li et al. (2008)
Pleurotus ostreatus	Fruit body	Methanolic	Tsai et al. (2008)

The Authors

Agaricus blazei Murril also called *Agaricus brasiliensis* is native to southern Brazil, popularly known as "Himematsuke" in Japan, or "Cogumelo do Sol" in Brazil. Consumption has increased in Brazil, Japan, Korea, Canada and United States because of its medicinal properties. Mechanism studies have demonstrated that antitumor activities of *A. blazei* extracts can be related to induction of apoptosis, cell-cycle arrest and inhibition of tumor-induced, neovascularization, immunopotentiation and restoration of tumor-suppressed host immune system.

The exopolysaccharide produced by *A. brasiliensis* showed strong inhibition against Sarcoma 180. The complete regression ratio was 50% and the suppression ratio percentage was 72.19%. Exopolysaccharide was characterized a mannan-protein complex, with its molecular weight being 10^5–10^7 by gel filtration and contained small amounts of glucose, galactose and ribulose (Fan et al., 2007).

Kim et al. (2009) optimized the extraction of *Agaricus blazei* for isolation of bioactives components with antitumor effects. Extracts of mushroom was obtained with different polarities and solubilities. One fraction, extracted at 80°C using 70% water-ethanol (v/v) shoed tumor inhibitory activity against the human promyelotic leukemia cells *in vitro*.

However, has been reported fresh mushroom fed at high amounts can be carcinogenic in mice. Lee et al. (2008c) showed dietary intake of *A. blazei* Murril fed at 6250, 12,500 and 25,000 ppm for two years of dry powder appears to enhance survival in males and not appear to be carcinogenic in rats.

Grifolin is a natural, biologically active substance isolated from fresh fruiting bodies of the mushroom *Albatrellus confluens*. Studies showed that grifolin is able to inhibit the growth of some cancer cell lines *in vitro* by induction of apoptosis. Ye et al., (2007) showed that grifolin inhibits the proliferation of nasopharyngeal carcinoma cell line through G1 phase arrest, which mediated by regulation of G1-related protein.

Cordyceps militaris is the best-known and most frequently collected bug-killing *Cordyceps*, but there are dozens of "entomogenous" species in North America (Kuo et al., 2006).

Park et al. (2009) demonstrated that water extracts of *C. militaris* may increase mitochondrial dysfunction, and results in the activation of caspase-9, leading to the activation of caspase-3 target proteins. The caspase family proteins are known to be one of the key executioners of apoptosis. Water extracts of *C. militaris* induces apoptosis in human lung carcinoma cells.

Coriolus versicolor, also known as Yun-Zhi produce bioactive compounds. Hot water and ethanol extracts from *C. versicolor* demonstrated activities antitumoral to treatment of melanoma cells. According to analysis, the predominant compounds are terpenoids and polyphenols. The prevention of tumor growth was exerted through diverse mechanism including cell cycle suspend, induction of tumor cell death by apoptosis and secondary necrosis, together with stimulation of the anti-tumor activity of macrophages (Harhaji et al., 2008).

Peng et al. (2005) also studied *Coriolus versicolor* and showed inhibitory effect on the growth of Sarcoma 180 solid tumors. Anti-tumor activity of polysaccharopeptide resides in its anti-angiogenic properties, via a suppression of vascular endothelial growth factor gene expression, resulting in a deprivation of angiogenic stimulation to the tumor growth.

Ganoderma lucidum is known as "mushroom of immortality" because enhancing longevity. Researchers have also demonstrated that *G. lucidum* inhibits the migration of breast cancer cells and prostate cancer cells, suggesting is potency to reduce tumor invasiveness. Since integrins are the major cell surface adhesion molecules expressed by all cell types. Tests showed that incubation with *G. lucidum* polysaccharides reduced integrin expression. Integrins are composed of α and ß transmembrane subunits. Each α and ß combinations has its own binding specificity and signaling properties (Wu et al., 2006).

Ganoderma lucidum polysaccharides used dose-dependently treatment enhance catalase activity in the polysaccharides-treatment groups when compared to the ovarian cancer model (Youguo et al., 2009).

Fruiting body of *Ganoderma tsugae* is used to promote health and longevity in Orientals countries. Peng et al. (2003) related that results indicate that ß-D-galacto-α-D-mannan isolated from culture filtrate of *G. tsugae* mycelium also exhibited significant antitumor activities. In 2005, the same researchers demonstrated anti-tumor activities were observed in three polysaccharides fractions with inhibition ratio 50%.

Mushrooms have antitumoral properties is controversy, sometimes researchers describe collateral effects. Sadava et al. (2009) analyzed cytotoxicity of twelve species of *Ganoderma* and just four species showed non-cytotoxic effects. However, active *Ganoderma* extracts induced apoptosis.

Grifola frondosa, has been reported to posses many biologically active compounds. Especially, the antitumor and immune-stimulating activities of polysaccharide D-fraction, a branched ß-(1-6)-D-glucan isolated from the fruiting body. Most reports conformed that mushroom polysaccharides exerted their antitumor action via activation on the immune response of the host organism, and mushroom polysaccharides were regarded as biological response modifiers. Polysaccharides from different strains have different antitumor activities *in vitro*. The data suggest that the polysaccharides fractions from *G. frondosa* had selective antitumor activities on the different tumor cell lines (Cui et al., 2007).

Surenjav et al. (2006), studied lentinan, (1-3)-ß-D-glucan, a antitumor polysaccharide, has been isolated from the fruiting body of *Lentinus edodes*. The triple helical (1-3)-ß-D-glucan antitumor containing protein showed activities against the growth of Sarcoma 180. The triple helical conformation plays an important role in the enhancement of the antitumor activities. Data suggesting that the antitumor activity of polysaccharide is also related to their molecular weight and content of the bound protein.

Li et al. (2008a) describe antitumoral activity of *Hedysarum polybotrys* Hand.-Maz (HP). In China, is used in the treatment of diseases, such as cancer, glycemy and immunomodulatory, anti-aging, anti-oxidation activities. The α-(1→4) D-glucan showed that inhibit proliferation of human hepatocellular carcinoma and human gastric cancer.

Inonotus obliquus is a white rot fungus, called Chaga, is a medicinal mushroom that has been used in Siberian and Russia folk medicine to treat stomach discomforts. Extracts of *I. obliquus* are known to inhibit the growth and protein synthesis of tumor cells. Alpha-linked-fucoglucomannan isolated from cultivated mycelia of *I. obliquus* can inhibit tumor growth *in vivo*. The endopolysaccharide-mediated inhibition of tumor growth is apparently caused by an induced humoral immunity of the host defense system rather than by a direct cytotoxic effect against tumor cells (Kim et al., 2006).

Huang et al. (2007) studied *Poria cocos* called Fu Ling, it is collected between July and September in China. Polysaccharides fractions was tested and showed strong inhibition against leukemia cell proliferation at all concentrations. The three water-soluble fractions presented significantly high inhibition ratio of more 80% at concentration of 200 μg/ml.

Phellinus ignarius, an orange color mushroom is used to improve health and remedy various diseases, such as gastroenteric disorders, lymphatic diseases and cancer. Extracts from fruiting body of *Phellinus igniarus* inhibited the proliferation of SK-Hep 1 cells and RHE cells in a concentration-dependent manner, with IC_{50} values of 72 and 103 μg/ml (Song et al., 2008).

Pleurotus is important mushroom because have importance gastronomic, nutritional, commercially and medicinal properties. *Pleurotus citrinopileatus* is a widely used edible mushroom, delicious taste and rich in nutrients. Antitumor activities of lecitin from *P. citrinopileatus* are similar to those *Pleurotus ostreatus* lecitin. *P. citrinopileatus* lecitin is dimeric, like lecitins from mushrooms (Li et al., 2008b). Sarangi et al. (2006) demonstrated that two fractions of *Pleurotus ostreatus* can directly kill Sarcoma 180 cells *in vitro*. Cell-cell adhesion determines the polarity of cells and participates in the maintenance of the cell societies called tissues. Adhesion is generally reduced in human cancer cells. Reduce intercellular adhesion allows cancer cells to disobey the social order, resulting in the destruction of histological structure, which is the morphological hallmark of malignant tumors. Reduced intercellular adhesiveness is also indispensable for cancer invasion and metastasis. Tong et al. (2009) observed also antitumor activity against HeLa tumor cell *in vitro*, in a dose-dependent manner and exhibited lower cytotoxicity to human embryo kidney cells.

Leucopaxillus giganteus is enormous mushroom is often found growing in large fairy rings or arcs in woodland clearings. It is apparently widely distributed, but most common in the Pacific Northwest and Rocky Mountains (Kuo et al., 2006).

Ren et al. (2008) demonstrated that clitocine isolated from *L. giganteus* have proliferation inhibitory activity against HeLa cells in a dose-dependent manner by mechanism involved the induction of apoptosis.

Immunomodulador Activity

Molecules of macrofungi and secondary metabolites are known as bioactive compounds that belong to polysaccharides, glycoproteins, involved in the innate and adaptive immunity, resulting in the production of cytokines. The therapeutic effects of these compounds such as antitumor and anti-infective activity and suppression of autoimmune diseases have been associated in many cases with their immunomodulating effects (Table 7.7).

Table 7.7. Potential immunomodulatory activity.

	Immunomodulator Activity		
	Source	**Extraction/compound**	**References**
Agaricus blazei	Fruit body	Water soluble compounds	Kasai et al. (2004)
Agaricus blazei	Fruit body/mycelium	Biocompounds	Shimizu et al. (2002)
Coriolus versicolor	Submerged fermentation	Polysaccharide	Lee et al. (2006)
Ganoderma lucidum	-	Fractions	Ji et al. (2007)
Ganoderma lucidum	-	Polysaccharide	Zhu et al. (2007)

Table 7.7. *(Continued)*

	Immunomodulator Activity		
	Source	**Extraction/compound**	**References**
Grifola frondosa	Submerged fermentation	Polysaccharide	Yang et al. (2007)
Grifola frondosa	Submerged fermentation	Polysaccharide	Wu et al. (2006)
Inonotus obliquus	Submerged fermentation	Polysaccharide	Kim et al. (2006)
Lentinula edodes	Fruit body	Polysaccharide	Gu and Belury (2005)
Lentinula edodes	Fruit body	Polysaccharide	Zheng et al., 2005

The Authors

Compounds that are capable of interacting with the immune system to up regulate or down regulate specific aspects of the host response can be classified as immunomodulators or biologic response modifiers. These agents can be applied in treating and preventing diseases and illnesses and with regard to the increase in diseases involving immune dysfunction, such as compounds with remedy potential without side effects, pathogenic resistance or affecting normal cell division (Moradeli et al., 2007).

Innate immunity serves as an essential first line of defense against microbial pathogens and may also influence the nature of the subsequent adaptive immune response. Phagocytic cells, such as macrophages and neutrophils, play a key role in innate immunity because of their ability to recognize, ingest, and destroy many pathogens by oxidative and non-oxidative mechanisms. Bioactive polysaccharides and polysaccharide-protein complexes have been isolated from mushrooms (Table 7.7), yeast, algae, lichens and plants, and these compounds have attracted significant attention because of their immunomodulatory and antitumor effects (Xie et al., 2007).

Mushroom polymers have immunotherapeutic properties by facilitating growth inhibition and destruction of tumors cells. Fungal ß-glucans-induced immune responses are different in their actions in immune therapies based on supplementation of elements of the immune system and stimulating the immune system, can be option in treatment of diseases. This compounds are not synthesized by humans and inducing both innate and adaptive immune response (Chen and Seviour, 2007).

Proteoglycans and polysaccharide have high molecular mass, constituent of β-glucans, cannot penetrate cells, so the first step in the modulation of cellular activity is binding to immune cell receptors. The mechanism by which the innate immune system recognizes and responds to compounds of mushroom is complex and multifactorial process. After this activation and signaling is humoral- and cell-mediate immunity induction (Moradali et al., 2007).

Agaricus blazei is a medicinal mushroom originating from Brazilian subtropical regions and is produced on an industrial scale in some countries such as China, Japan and Brazil (Lima et al., 2008).

Extracts of *Agaricus blazei* from the fruiting body and the mycelium were effective in activation of the human complement pathway. Both bioactive compounds have been demonstrated to be potent activators of the complement system in human serum in a dose and time dependent manner (Shimizu et al., 2002).

Kasai et al. (2004) analyzed *Agaricus blazei* fraction induced expression of IL-12, a cytokine known to be a critical regulator of cellular immune responses. According to Kimura et al. (2006), supplementation of *A. blazei* was effective in the activation of enzymes related to energetic metabolism in leukocytes of calves. *A. blazei* extract was water-soluble and easy to deal with as a food additive.

Lentinan, schizophyllan, and krestin have been accepted as immunoceuticals in several oriental countries. Increase of natural killer cells, cytotoxic T lymphocytes and delayed type hypersensitivity responses against tumor antigen were observed after administration of lentinan. *Lentinula edodes* was claimed to have a range of health benefits. The concentration of TNF-α, IFN-ν in serum increased significantly in the polysaccharide groups compared with the model control group. Also, polysaccharide of *L. edodes* increased oxide nitric production in peritoneal macrophages and catalase activity of macrophage (Zheng et al., 2005).

Coriolus versicolor, known as Yun Zhi, has been used in China for treatment of cancer and immunodeficient. Polysaccharopeptides produced for *C. versicolor* can stimulate cytokines production but also demonstrated that a critical time for culture harvesting is essential for obtaining optimal bioactivities of the fungi (Lee et al., 2006).

Ganoderma lucidum is a Chinese medicinal fungus, which has been clinically used in East Asia and is given considerable attention for treatment for various diseases. Medicinal functions have been assigned to crude extracts and isolated components of *G. lucidum.* The potential immunomodulating activity is the capacity of a particular substance to influence specific immune functions such as activating individual components of the immune system and promoting cytokine synthesis. Ji et al. (2007) showed that extracts of *G. lucidum* activated mouse macrophages in a dose dependent manner, increased the levels of IL-1, IL-12p35 and IL-12p40 gene expression, and significantly enhanced oxide nitric production. These immunomodulatory functions suggest that *G. lucidum* may also interfere with the growth of certain tumors.

Bao et al. (2001) demonstrated that (1-3)-D-Glucans of *Ganoderma lucidum* have immunomodulating and antitumor activities. The structural and physicochemical properties and lymphocyte proliferation activity of all samples varied with the functionalized groups and the degree of substitution. The results of immunological assays indicated that some modified derivatives had stimulating effects on lymphocyte proliferation and antibody production and the introduction of carboxymethyl group with low degree of substitution was the best choice on the improvement of the immunostimulating activity.

Ganoderma lucidum mycelia stimulated moderate levels of TNF, IL-6 and IFN-release in human whole blood and moderately stimulate cytokine production without potentiating oxide nitric release. The ineffectiveness in inducing oxide nitric release by *G. lucidum* mycelia indicates that the compositions and structures of glucan in mycelia and fruiting body may be different, and this might result in enhancing innate immune response through different receptors or pathways (Kuo et al., 2006).

Ganoderma lucidum polysaccharides enhanced the activity of immunological effectors cells in immunosuppressed mice and promoted phagocytosis and cytotoxicity of macrophages. The above beneficial effects induced by the low-dose of polysaccharide treatment did not result in any side effects (Zhu et al., 2007).

Polysaccharides obtained from fermented and fruiting body of *Grifola frondosa* have demonstrated many interesting biological activities. Exopolymers fractions of *G. frondosa* can be enhancers of innate response and considered as potent materials for immune system (Shih et. al., 2008; Yang et al., 2007).

Grifolan, polysaccharide of *Grifola frondosa,* showed that hot water-soluble fractions (polysaccharides) of mycelia from submerged fermentation can effectively induce innate immunity and therefore enhance pro-inflammatory cytokine release, phagocytosis, and Natural Killer cytotoxicity activity *in vitro* (Wu et al., 2006).

Inonotus obliquus is a white rot fungus widely distributed over Europe, Asia, and North America. The polysaccharide yield of species of *Inonotus* increased proportionally with an increasing cell mass of the fermentation broth. However, the immunostimulating activities were not proportional to the corresponding polysaccharide yield. High specific activities of endopolysaccharide were obtained during both the late lag and the late stationary phases, but not during the active cell growth phase, indicating that the polysaccharide activity is probably closely related to cell age. During the late lag phase, low total activities were obtained due to low cell masses in spite of high specific activities. The specific activity of endopolysaccharide at late lag phase appeared to be highly similar to the activity at late stationary phase. The endopolysaccharide of *I. obliquus* showed much higher splenic cell activities than the corresponding exopolysaccharide (Kim et al., 2006).

Anti-angiogenic Activity

Angiogenesis can be characterized as an integrate set of cellular, biochemical and molecular processes in which new blood vessels are formed from pre-existing vessels. This occurs physiologically during reproductive and developmental processes as well as during the late phases of wound healing following tissue damage (Contois et al., 2009). Blood vessels run through every organ in the body (except cornea and cartilage), assuring metabolic homeostasis by supplying oxygen and nutrients and removing waste products. Therefore, angiogenesis is known to be essential in several physiologic processes, such as organ growth and development, wound healing and post-ischemic tissue repair. However, inappropriate or aberrant angiogenesis contributes to the development and progression of various pathological conditions including tumor growth and metastasis, diabetic retinopathy, cardiovascular diseases, inflammatory disease and psoriasis. Angiogenesis can be separated into several main steps, such as degradation of the basement membrane of exiting blood vessels, migration, proliferation and rearrangement of endothelial cells, and formation of new blood vessels (Makrilia et al., 2009; Ramjaum and Hodivala-Dilke, 2009; Ribatti, 2009).

This switch clearly involves more than simple up-regulation of angiogenic activity and is known to be the result of net balance between positive and negative regulators. There are three particularly important stimulators of angiogenesis: (i) vascular endothelial growth factor (ii) fibroblast growth factor; (iii) angiopoetins; between many others, like platelet derived growth factor, epidermal growth factor, ephrins;

transforming growth factors alpha and beta, interleukins, chemokines, and small molecules, such as sphingosine 1-phosphate, that are known to promote cell proliferation, survival and differentiation of endothelial cell (Duarte et al., 2007; Jung et al., 2008; Stupack et al., 1999).

Anti-angiogenesis strategies are based on inhibition of endothelial cell proliferation, interference with endothelial cell adhesion and migration, and interference with metalloproteases. Down-regulation of angiogenesis has been considered to be advantageous for the prevention of tumors (Bhat and Singh, 2008; Jung et al., 2008; Song et al., 2003).

In Taiwan, the mushroom *Antrodia cinnamomea* is known as "niu-cha-ku" or "chang-chih" and produced triterpene acids, steroid acids and polysaccharides with biological activities. The fraction >100 kDa of polysaccharide from *A. cinnamomea* showed potential anti-angiogenic *in vivo* and *ex vivo* indirectly by immunomodulation (Table 7.8) (Yang et al., 2009).

Table 7.8. Anti-angiogenesis activity of mushrooms.

Mushroom	Anti-angiogenesis Activity		
	Source	**Extraction/compound**	**References**
Antrodia cinnamomea	Mycelia	Hot water	Yang et al. (2009)
Ganoderma tsugae	Fruit body	Methanol extracts	Hsu et al. (2009)
Grifola frondosa	Fruit body	Water extracts	Lee et al. (2008b)
Fomitopsis pinicola	Fruit body	Polysaccharides	Cheng et al. (2008)
Phellinus linteus	Fruit body	Ethanolic extract	Song et al. (2003)

The Authors

Ganoderma sp. contains numerous bioactive natural components and the two categories of those are polysaccharides and tripernoids, both of them are potent inhibitors of *in vitro* and *in vivo* tumor growth. The epidermal growth factor receptor activation is often linked with angiogenesis. Methanol extracts of *Ganoderma tsugae* showed antiangiogenic effects on the cancer cells by the downregulation of vascular endothelial growth factor (Hsu et al., 2009).

Fomitopsis pinicola is marketed as a tea and food supplement. Chemical compounds found in *F. pinicola* include steroids, sesquiterpenes, lanostane tripertenoids and triterpene glycosides. Ethanolic extracts and polysaccharides of *F. pinicola* were effective for the anti-angiogenesis at 10 μg/ml concentration and showed no toxicity up to concentration of 1 mg/ml (Cheng et al., 2008).

Grifola frondosa demonstrated anti-angiogenic activity by inhibit vascular endothelial growth factor induced angiogenesis *in vivo* and *in vitro.* Vascular endothelial growth factor is the most angiogenic factor associated with inflammatory diseases and cancer (Lee et al., 2008b). Ethanolic extracts from *Phellinus linteus* presented anti-angiogenic activity and can be used as adjuvant chemotherapy for the treatment of cancer (Song et al., 2003)

Hypoglycemic Activity

The WHO describes diabetes as a chronic condition that occurs when the pancreas does not produce enough insulin or when the body cannot effectively use the insulin it produced. Hyperglycemia and other related disturbances in the body's metabolism can lead to serious damage to many body's systems, especially to the nerves and blood vessels. Diabetes is a life threatening condition. WHO indicates that worldwide almost 3 million deaths per year are attributable to diabetes.

Roglic et al. (2005) describe that excess mortality attributable to diabetes accounted for 2–3% of deaths in poorest countries and over 8% in the U.S.A., Canada, and the Middle East. In people 35–64 years old, 6–27% of deaths were attributable to diabetes. Functional foods and naturals compounds have become a popular approach to prevent occurrence of diabetes mellitus. Several mushroom species have been described to have anti-diabetic properties, because were found compounds as fibers source, polysaccharides and other biological activities (Table 7.9).

Table 7.9. Potential biological of mushrooms.

	Activity Hypocholesterolemic		
Auricularia auricular	Fruit body	Powder	Cheung (1996)
Pleurotus ostreatus	Mycelia	Proteo-glucan	Sarangi et al. (2006)
Pleurotus ostreatus	Fruit body	Dried Fruit body	Bobek et al. (1998)
Tremella fuciformes	Fruit body	Powder	Cheung (1996)
	Activity Hypoglycemic		
Cordyceps sinensis	Fruit body	Polysaccharide	Li et al. (2006)
Coprinus comatus	Submerged fermentation	Vanadium	Han et al. (2006)
Marasmius androsaceus	Mycelia	Ethanol extracts	Zhang et al. (2009)
Phellinus baumii	Submerged fermentation	Polysaccharide	Hwang et al. (2005)
Sparassis crispa	Fruit body	Powder	Kwon et al. (2009)
Tremella mesenterica	Submerged fermentation and fruit body	Fruit body Water extracts	Lo et al. (2006)
	Activity Anti-hypertensive		
Lyophyllum decastes (Sing.)	Fruit body	Powder and water extracts	Kokean et al. (2005)
Pleurotus nebrodensis	Fruit body	Polysaccharide and water extracts	Miyazawa et al. (2008)
	Biosynthesis of collagen		
Grifola frondosa	Submerged Fermentation	Polysaccharides	Lee et al. (2003)
	Activity Mitogenic		
Pleurotus citrinopileatus	Fruit body	Lectin	Li et al. (2008)
	Activity Prebiotic		
Pleurotus ostreatus	Fruit body	Glucans	Synytsya et al. (2009)

The Authors

Lima et al. (2008) the diet supplemented with exopolysaccharides of *Agaricus brasiliensis* provided during 8 weeks to the mice produced a reduction in glucose plasma concentration around 22%. It has been also demonstrated in another study that the supplementation with β-glucans and oligosaccharides obtained from the fruiting body of *A. brasiliensis* caused reduction in the glucose serum concentration in rats. Mushroom could have an anti-diabetic activity by promoting the insulin release by the Langerhans cells in the pancreas. The total cholesterol ratio in the mice was reduced around 27% with *A. brasiliensis* exopolysaccharide supplementation. Fruiting body biomass and the mushroom polysaccharides have significant anti-hyperglycemic activity and the abilities to increase glucose metabolism and insulin secretion in type 2 diabetes mellitus. However, the mechanisms of action for the exopolysaccharide of *A. brasiliensis* on cholesterol and glucose metabolism are still unknown. The effect of *A. brasiliensis* on the enzyme activities was considered to be sustained for at least two to three months after the supplementation in the calves, although the absorption of *A. brasiliensis* was considered satisfactory.

Vanadium compounds have the ability to imitate action of insulin. Oral administration of inorganic vanadium salts, have shown anti-diabetic activity *in vitro*, *in vivo* and even in patients. Vanadium at lower doses (0.18 mg/kg/d) was absorbed by fermented mushroom of *Coprinus comatus*, which is one rare edible fungus that is able to absorb and accumulate trace elements. *C. comatus* is a mushroom claimed to benefit glycemic control in diabetes and others properties. *C. comatus*, on a dry weight basis, contains, on the average, 58.8% carbohydrate, 25.4% protein and 3.3% fats, with the rest constituted of minerals. It indicates that *C. comatus* could supplement nutrients to the mice as well as lower blood glucose of hyperglycemic mice. *C. comatus* has the ability to take up and accumulate trace metals. However, the toxicity associated with vanadium limits is therapeutic efficacy (Han et al., 2006).

Tremella mesenterica contains up 20% of polysaccharide glucuronoxylomannan in the fruiting bodies, is a popular, edible and medicinal mushroom in orient. Fruiting bodies, submerged fermentation and the acid polysaccharide of *T. mesenterica* have significant anti-hyperglycemic activity. Consumption of *T. mesenterica* by rats had significantly improved short-and long-term glycemic responses, as evidence by significantly decreased blood glucose concentrations in oral glucose tolerance test and serum fructosamine concentrations, respectively. This mushroom has potential anti-hyperglicemic functional food or nutraceuticals for diabetes with daily ingestion of 1 g/Kg of fruiting bodies, biomass and glucuronoxylomannan. Also, *T. mesenterica* has the ability to increase the insulin sensitivity, instead of increasing the insulin secretion in normal rats. Consumption of *T. mesenterica* may act functional food in improving the short and long-term glycemic control in persons with high risk of diabetes or type 2 diabetes mellitus, not in persons with type 1 diabete mellitus (Lo et al., 2006).

Phellinus baumii is a mushroom used as a folk medicine for a variety of human diseases in several Asian countries. The plasma glucose level in the exopolisaccharide-fed rats was substantially reduced by 52.3% as compared to the diabetic rats, which is the highest hypoglycemic effect among mushroom-derived. The activities of alanine aminotransferase and asparate aminotransferase were significantly decreased by administration of *P. baumii* exopolysaccharide, thereby exhibiting a remedial role in liver function. *P. baumii* of exopolysaccharide administration led to the diabetogenic effect and significantly reduced the degree of diabetes. Oral administration of *P. baumii* of exopolysaccharide may have a potential benefit in preventing diabetes, since pancreatic damage induced by environmental chemicals and other factors is a cause of diabetes (Hwang et al., 2005).

Exopolysaccharides of *Ganoderma lucidum* (Lingzhi) have hypoglycemic effects. Studies showed that treatment with water extract of *Ganoderma lucidum* for 4-week oral gavages, 0.3 g/kg for consumption, lowered the plasma glucose level in mice. Phosphoenolpyruvate carboxykinase is a hepatic enzyme which is important in the regulation of gluconeogenesis. Inhibition of the hepatic phosphoenolpyruvate carboxykinase reduced blood glucose levels and improved glucose tolerance together with a decreased circulating free fatty acid and triacylglycerol levels in the diabetic mice. *G. lucidum* consumption caused a marked suppression of the hepatic phosphoenolpyruvate carboxykinase gene expression with a concomitant reduction of the serum glucose levels in mice (Seto et al., 2009).

Cordyceps, one of the most valued traditional Chinese medicines, consists of the dried fungus *Cordyceps sinensis* growing on caterpillar. Polysaccharide from *Cordyceps* protects against free radical induced neuronal cell toxicity. Polysaccharides of *Cordyceps* produced a drop in blood glucose level in both normal and diabetic animals, at doses of higher than 200 mg/kg body wt. Hypoglycemic effect is possibly because of the increase in blood insulin level, which may be due to the induced insulin release from the residual pancreatic cells and/or reduced insulin metabolism in body by polysaccharide (Leung et al., 2009; Li et al., 2006).

Mechanisms which contribute to the formation of free radicals in diabetes include non-enzymatic and auto-oxidative glycosylation, metabolic stress resulting from changes in energy metabolism, levels of inflammatory mediators, and the status of antioxidant defense. Selective damage of islet cells in the pancreas may be one of the pathological mechanisms for Type I diabetes. Antioxidants could prevent the development of diabetes. Polysaccharide also has a strong antioxidant property which can protect cultured rat pheochromocytoma PC12 cells from being damaged by hydrogen peroxide. This antioxidant activity of polysaccharide may also play a protective role in the development of diabetes (Li et al., 2006).

Kwon et al. (2009), described consequences caused for diabetes, such as difficult proliferation of cells, decreased collagen production, decreased chemotaxis and phagocytosis, reduction in the levels of growth factors, and the inhibition of fibroblast proliferation. They tested *Sparassis crispa*, an edible medicinal mushroom consumed in China and Japan. In experiments, the diabetes was induced and was accompanied by diabetic symptoms such as weight loss, polyuria, hyperglycemia, and neuroendocrine dysfunction. The oral administration of *S. crispa* increased migration of macrophages and fibroblasts, collagen regeneration, and epithelialization under hyperglycemic conditions.

Anti-Hypertensive Activity

According to WHO, cardiovascular diseases include coronary heart disease (heart attacks), cerebrovascular disease, raised blood pressure (hypertension), peripheral artery disease, rheumatic heart disease, congenital heart disease and heart failure. The major causes of cardiovascular disease are tobacco use, physical inactivity, and an unhealthy diet. alcohols (IC_{50}:16.4 mM) and were the most effective inhibitor of Angiotensin I converting enzyme. Although sugar alcohol might prevent an increased in blood pressure by mechanism such as osmotic diuretic effect, more studies are necessary to explain the mechanism *in vivo* (Hagiwara et al., 2005).

Pleurotus nebrodensis is native from Southern Europe, Central Asia and China, and have been shown to prevent hypertension. Compounds of *P. nebrodensis* were administered orally, and antihypertensive actions were measuring blood pressure. Two hours after administration, the blood pressure decreased around 85% and increased gradually after 6 until 48 hours pos-administration (Miyazawa et al., 2008).

Marasmius androsaceus, a traditional chinese mushroom, is usually used in tendon relaxation, pain alleviation and anti-hypertension. The 3,3,5,5 Tetramethyl-4-piperidone is an active compound prepared from *Marasmius androsaceus* but, the action unclear. Study showed that 3,3,5,5 tetramethyl-4-piperidone have effects reducing blood pressure in anesthetic tested in rats and cats. It can inhibit the automatic rhythmic contraction of ileum section in guinea pig. It also can inhibit the concentration of rabbit aorta smooth muscle caused by adrenalin. 3,3,5,5 tetramethyl-4-piperidone has a simple structure with low molecular weight, which is suitable to serve as leading compound. This is an antihypertensive compound, and the effect is partially related to ganglionic blocking (Zhang et al., 2009).

Prebiotics

Prebiotic is as "selectively fermented ingredients that allow specific changes, both in the composition and/or activity in the gastrointestinal microbiota that confers benefits upon host well-being and health." The effect of a prebiotic is indirect because it selectively feeds one or a limited number of microorganisms thus causing a selective modification of the host's intestinal microflora. Intestinal bacteria leads towards a consideration of factors that may influence the flora composition in a manner than can impact upon health (Reid, 2008; Vasiljevic and Shah, 2008). The criteria for consider prebiotic are: resistance to the upper gut tract, fermentation by intestinal microbiota, beneficial to the host health, selective stimulation of probiotics, stability to food processing treatments (Wang, 2009).

Ingestion of prebiotic was believed to enhance immune function, improve colonic integrity, decrease incidence and duration of intestinal infections, down-regulated allergic response as well as improve digestion and elimination of faeces (Aida et al., 2009).

Mushrooms are consumed as a delicacy, and particularly for their specific aroma and texture. Digestibility and bioavailability of mushroom constituents have been missing from the knowledge of mushroom nutritional value. The dry matter of mushroom fruit bodies is about 5–15%, they have a low fat content and contain 19–35%

proteins. The content of carbohydrates, which are mainly present as polysaccharides or glycoproteins, ranges 50–90%. Most abundant mushroom polysaccharides are chitin, hemicelluloses, ß- and α-glucans, mannans, xylans and galactans. Mushroom polysaccharides are present as linear and branched glucans with different types of glycosidic linkages, such as (1-3), (1-6) - ß-glucans and (1-3) α-glucans, but some are heteroglicans containing glucorinic acid, xylose, galactose, mannose, arabinose or ribose. Like polysaccharides originated from other food products, they contribute to the digestion process as soluble or insoluble dietary fibers depending on their molecular structure and conformation. ß-glucans are recognized as immunological activators and are effective in treating diseases like cancer, diabetes, hypercholesterolemia and others. Mushroom seem to be a potential in probiotics candidate for prebiotics as it contains carbohydrates like chitin, hemicelulose, ß- and α-glucans, mannans, xylans and galactans (Table 7.9). Chitin is a water-insoluble structural polysaccharide, accounting for up to 80–90% of dry matter in mushroom cell walls. A high proportion of indigestible chitin apparently limits availability of other components (Aida et al., 2009; Kalac, 2009; Synytsya et al., 2009).

Synytsya et al. (2009) studied extract of *Pleurotus ostreatus* and *Pleurotus eryngii* with potential prebiotic activity. Specific soluble glucans were isolated from mushrooms by boiling water and alkali extraction. Potential prebiotic activity of extracts aqueous and alkali extracts was testing using nine probiotic strains of *Lactobacillus*, *Bifidobacterium* and *Enterococcus*. The utilization of both extracts by different manner affirms different chemical structure of polysaccharides, such as prebiotics. Extracts of *Pleurotus* can be used for symbiotic construction with select probiotic strains.

CONCLUSION

Mushrooms are traditionally used in oriental countries like food and several diseases treatment. The science widens opportunity of progress in the development of new diseases treatments, by discovering the natural compounds.

KEYWORDS

- ***Agaricus blazei***
- ***Agrocybe cylindracea***
- ***Boletus edulis***
- ***Pleurotus ostreatus***
- **Tocopherols**

Chapter 8

Ferulic Acid Content and Antioxidant Capacity in Pinto Bean (*Phaseolus vulgaris* L.) Varieties

Agustín Rascyn-Chu, Karla Escárcega-Loya, Guillermina GarcHa-Sánchez, Elizabeth Carvajal-Millán, Alejandro Romo-Chacyn, and Jorge Márquez-Escalantle

INTRODUCTION

The ferulic acid content and antioxidant activity of two Mexican beans varieties was investigated. *Pinto Villa* beans presented a higher ferulic acid content (137 mg/100g) and antioxidant capacity (33 mmoles Fe^{++}) in comparison to *Pinto Saltillo* variety (47 mg/100g and 22 mmoles Fe^{++}, respectively). For *Pinto Villa*, cooking process reduced ferulic acid content and antioxidant activity by 38 and 55%, respectively, while *Pinto Saltillo* showed a decrease of 15 and 64%, respectively. These results indicate that *Pinto Villa* bean variety could be a better antioxidant source in spite of a less appealing appearance to consumers.

Functional food (FF) has neither a precise nor a universal definition; rather, our understanding of FF is based on its beneficial effects on health promotion and/or disease prevention. Despite its broad conceptual definition, the common notion is that naturally occurring active components in foods help to define whether or not they are FF (Verschuren, 2002). Several of the ingredients reported as functional components, such as catechins and ferulic acid, are found in beans. Bean consumption has been related to reduce the risk of diabetes, cancer, and cardiovascular diseases. The potential health benefits of beans have been attributed to the presence of micronutrients, such as phenolic compounds, that possess antioxidant properties. Ferulic acid is the predominant phenolic compound in beans (Luthria and Pastor-Corrales, 2006). Phenolics can act as free radical terminators, chelators of metal catalysts, or singlet oxygen quenchers (Shahidi and Wanasundra, 1992). Consumption of free radicals and oxidation products may be a risk factor for cancer and cardiovascular disease, and dietary phenolics may have health benefits (Huang et al., 1992).

Beans play an important role in human nutrition since they are rich in protein, fiber, certain minerals, and vitamins. *Pinto* beans (*Phaseolus vulgaris* L.) have traditionally been a staple food for the people in Northern Mexico. Nevertheless, beans are used in a lesser extent as foreign cultures influence consumption. Furthermore, new *Pinto* beans varieties as *Pinto Saltillo* are being introduced into market. These new varieties are resistant to storage being more interesting to producers and are colorless varieties being more appealing to consumers in comparison to classic *Pinto* varieties as *Pinto Villa*. Nevertheless, the effect of these new features on antioxidant capacity has not been yet assessed. The purpose of this work was to study the ferulic acid content and its reflect on antioxidant activity of a relatively new Pinto Mexican bean variety and one of its parental lines, and how these values are affected in bean samples after the cooking process.

MATERIALS AND METHODS

Materials

Pinto Villa and *Pinto Saltillo* Mexican bean (*Phaseolus vulgaris* L.) varieties were supplied by a commercial proveyor in Northern Mexico. All chemicals were reagent pure grade.

Cooking Process

Pinto beans were cooked in a pressure casserole during one hour. A bean:water ratio of 1:2 (w/v) was used.

Ferulic Acid Content

Ferulic acid contents in the samples were quantified by HPLC after deesterification step as described by Vansteenkiste et al., (2004). 100 mg of sample was allowed to react with 1 ml of 2N NaOH for two hours in the dark at 35°C under argon. After adding 3,4,5-trimethoxy-trans-cinnamic acid (TMCA, internal standard, 10 μg), the pH was adjusted to 2.0 ± 0.2 with 4N HCl. Phenolics were extracted twice with diethyl ether, and evaporated at 30°C under argon. The dried extracts were solubilized in 0.50 ml methanol/water/acetic acid (40/59/01), filtered (0.45 μm) and injected (20 μl) into HPLC using a Supelcosil LC-18-DB (250 ′ 4.6 mm, particule size 5μl) preceded by a guard column (supelcosil™, 250 mm) (Supelco, Inc., Bellefont, PA, USA) column. Detection was by UV absorbance at 280 nm. Isocratic elution was performed using methanol/water/acetic acid (40/59/01) at 0.6 ml/min at 35°C. A Varian, a variable wavelength UV-Vis detector 9050 (Varian, St. Helens, Australia) was used to record the ferulic acid spectra. A Star Chromatography Workstation system control version 5.50 was used.

Antioxidant Capacity

Antioxidant capacity was performed following the FRAP method (Benzie and Strain, 1996). One g of sample was mixed with 9 ml of 100 mm phosphate buffer (pH 7.6 containing 0.1 mm EDTA) during 30 min at 25°C. The mixture was centrifuged (15,000 rpm, 15 min, 25°C) and supernatant was used for the measurements. The FRAP reagent was prepared by mixing 30 mm acetate buffer (pH 3.6), 40 mm TPTZ (2, 4, 6-tripyridyl-s-triazine) and 20 mm $FeCl_3$ in a ratio of 10:1:1 (v/v). A total of 100 μl of sample supernatant, 300 μl of distilled water and 3 ml of FRAP reagent were pipetted into test tubes and incubated at 37°C for 4 min. Each sample was run in triplicate. Aqueous solution of known Fe (II) concentration was used for calibration (in a range of 100–1000 μM). A Cary 1E Varian UV-Visible Spectrophotometer (Varian, St. Helens, Australia) was used to measure the absorbance at 593 nm.

Color

Ground samples were analyzed with a color difference meter (model CR-300, Minolta, Japan). Lightness (L), redness (a), and yellowness (b) values were recorded. Each value was an average of ten different independent measurements. The browning index (BI) was calculated from these measurements as reported by Mohammadi et al., (2008).

Statistical Analysis

Data were analyzed with an ANOVA. Means separation was according to Tukey ($p \leq 0.05$) (MINITAB 13 version).

RESULTS AND DISCUSSION

Ferulic acid content in raw and cooked *Pinto Villa* and *Pinto Saltillo* beans varieties is presented in Table 8.1. Luthria and Pastor-Corrales (2006) reported a total phenolic acid content varying between 19.1–48.3 mg/100 g of raw beans, ferulic acid being the most abundant phenolic acid. The ferulic acid content in *Pinto Villa* variety found in the present study was 137 mg/100 g, higher to that reported by these authors, while *Pinto Saltillo* value is in the range 47 mg/100 g, a value close to the value found for cvar. Villa after cooking (58 mg AF/100 g). Ferulic acid is heat labile, and *Pinto Villa* and *Pinto Saltillo* retained 42 and 13% of the ferulic acid content after the cooking process we used, respectively (Table 8.1). Nevertheless in a previous study, Luthria and Pastor-Corrales (2006) found that 83% of the total phenolic acids were retained in bean samples during the cooking process. A 200 g serving of cooked *Pinto Villa* and *Pinto Saltillo* beans varieties provides 114 and 12 mg of ferulic acid, respectively, which are 12 and 1.3% of the IDR for adults. Baublis et al., (2000) found that ferulic acid had a strong antioxidant capacity. The antioxidant activity in these beans is similar to that of ascorbic acid, α-tocopherol or uric acid, which is around 2.0 mmoles Fe^{++} (Benzie and Strain, 1996). The health benefits of ferulic acid are mainly due to its antioxidant activity, free radical scavenging, and chelating of redox active metal ions. The total polyphenol intake can be calculated from the polyphenol contents in food and food consumption tables. Kühnau (1976) determined a flavonoid intake in the United States of ~1 g/d. In raw samples, *Pinto Villa* beans presented a higher antioxidant capacity (33 mmoles Fe^{++}) in comparison to *Pinto Saltillo* variety (22.5 mmoles Fe^{++}). For *Pinto Villa* cooking process reduced antioxidant activity in 46%, while *Pinto Saltillo* showed a decrease of 35.54% (Fig. 8.1). Even though, the remnant antioxidant activity is higher and evidently more desirable for *Pinto Villa*, than *Pinto Saltillo*. Up to this point of research, the evidence gathered, points out the higher value for phenolic and total antioxidant activity for the *Pinto Villa* cvar.

Table 8.1. Ferulic acid content in *Pinto Villa* and *Pinto Saltillo* beans varieties before and after cooking.

	Ferulic acid (mg AF/100 g db)	
Bean variety	*Raw*	*Cooked*
Pinto Villa	137	57
Pinto Saltillo	47	6

Mean values in the same column with different letters are significantly different ($p \leq 0.05$).

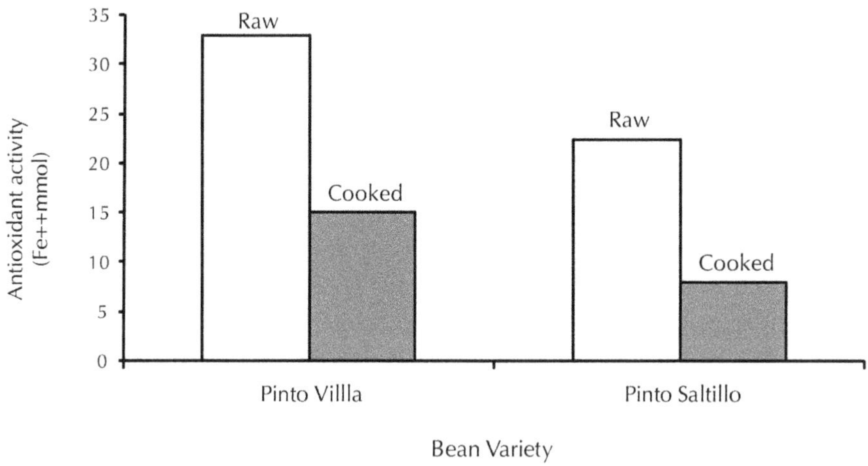

Figure 8.1. Antioxidant capacity of *Pinto Villa* and *Pinto Saltillo* beans varieties before and after cooking.

Interestingly, the actual tendency of consumption is deriving toward the production and consumption of *Pinto Saltillo* cvar based on color differences. In raw and cooked samples, lower color L* value was found in *Pinto Villa* bean in comparison to *Pinto Saltillo* bean variety, and even darker values for cooked samples of *P. Villa* cvar.(Table 8.2). Also the BI values calculated from L*, a*, and b*, measurements, point out the fact that cvar. *Saltillo* has a lighter appearance. The visual evidence is confirmed by the statistical difference within values apparently close. There are differences either between raw and cooked, or between cvars. This behavior on color is due to the lack of phenolic content as previous measurements indicate, therefore a lesser antioxidant activity. Eventhough the clearer appearance of *Pinto Saltillo* seems to be more appealing to some consumers in comparison to *Pinto Villa*; information regarding features like antioxidant properties and its health implications would eventually impact the consumption tendencies. The lighter color and less browning of *Pinto Saltillo* is not related to antioxidant agents present, but the opposite. The fact that cvar *Saltillo* is drought resistant should take into account the latter evidence as a reminder that nutritional features concerning the consumer should be considered when it comes to cvar improvement.

Table 8.2. *Pinto Villa* and *Pinto Saltillo* beans varieties physical characteristics.

Bean variety	Luminosity (L)		Browning Index (BI)	
	Raw	Cooked	Raw	Cooked
Pinto Villa	86.7 b	68.4 b	11.4 ± 0.19 a, a	36.7 ± 0.44 a, b
Pinto Saltillo	88.5 a	75.9 a	10.5 ± 0.38 b, a	32.6 ± 0.21 b, b

Note: Mean values in the same column with different letters are significantly different ($p \leq 0.05$).

CONCLUSION

Pinto Villa beans presented a higher ferulic acid content and antioxidant capacity in comparison to *Pinto Saltillo* variety. In this bean variety, 62 and 45% of the ferulic acid content and the antioxidant capacity, respectively, were retained in bean samples during the cooking process. These results showed that *Pinto Villa* bean variety could be a better antioxidant source in spite of a less appealing appearance to consumers and a less resistance to drought and/or storage.

FUTURE CONSIDERATIONS

Pinto Villa beans have traditionally played an important role for the people in Northern Mexico, since they are rich in nutrients. This bean variety continues to be investigated as a source of ferulic acid and other antioxidants. Nevertheless, *Pinto Villa* presents a high susceptibility to develop the hard-to-cook phenomenon. This defect results in economic losses, either through the rejection of beans by consumers for its poor texture or due to the need for increased energy required for cooking. Our evidences are not conclusive for phenolics/protein content ratio and their interaction. Therefore, additional studies will be required in order to understand the relationship between hard-to-cook phenomenon and the changes in the phenolic compounds of the grain, as related to the other components, such as proteins, phytates, and minerals. Further research on cell wall composition and physical properties should be addressed for eventually understand function, structure, and hard-to-cook features in beans.

ACKNOWLEDGMENTS

This research was supported by FOMIX CONACYT-Chihuahua (grant Chih-2006-C02-59228 to A. Rascon-Chu, PhD). The authors are pleased to acknowledge Jorge Márquez-Escalante and Ana Luisa Martínez-López for their technical assistance.

KEYWORDS

- **Antioxidant capacity**
- **Beans**
- **Ferulic acid**
- ***Pinto Saltillo***
- ***Pinto Villa***

References

1

Aguedo, M., Ly, M. H., Belo, I., Teixeira, J. A., Belin, J. M., and Waché, Y. (2004). The use of enzymes and microorganisms for the production of aroma compounds from lipids. *Food Technology Biotechnology*, **42**, 327–336.

Berger, R. G. (2009). Biotechnology of flavours—The next generation. *Biotechnology Letters*, Published on-line: July 16, 2009.

Carvalho, P. O., Calafatti, S. A., Marassi, M., Silva, D. M., Contesini, F. J., Bizaco, R., and Macedo, G. A. (2005). Potencial de biocatálise enantioseletiva de lipases microbianas. *Quimica Nova*, **28**, 614–621.

Christen, P. and López-Munguia, A. (1994). Enzymes and food flavor—A review. *Food Biotechnology*, **6**, 167–190.

Couto, S. R. and Sanromán, M. A. (2006). Application of solid-state fermentation to food industry—A review. *Journal of Food Engineering*, **76**, 291–302.

Ewis, H. E., Abdelal, A. T., and Lu, C. D. (2004). Molecular cloning and characterization of two thermostable carboxylesterases from Geobacillus stearothermophilus. *Gene*, **329**, 187–195.

Ghorai, S., Banik, S. P., Verma, D., Chowdhury, S., Mukherjee, S., and Khowala, S. (2009). Fungal biotechnology in food and feed processing. *Food Research International*, **42**, 577–587.

Hasan, F., Shah, A. A., and Hameed, A. (2005). Industrial applications of microbial lipases. *Enzyme and Microbial Technology*, **39**, 235–251.

Jayani, R. S., Saxena, S., and Gupta, R. (2005). Microbial pectinolytic enzymes: A review. *Process Biochemistry*, **40**, 2931–2944.

Kashima, A., Iijima, M., Nakano, T., Tayama, K., Koizumi, Y., Udaka, S., and Yanagida, F. (2000). Role of intracellular esterases in the production of esters by *Acetobacter pasteurianus. Journal of Biosciense and Bioengineering*, **89,** 81–83.

Kirk, O., Borchert, T. V., and Fuglsang, C. C. (2002). Industrial enzyme applications. *Current Opinion in Biotechnology*, **13**, 345–351.

Krishna, S. H., Manohar, B., Divakar, S., Prapulla, S. G., and Karanth, N. G. (2000). Optimization of isoamyl acetate production by using immobilized lipase from *Mucor miehei* by response surface methodology. *Enzyme and Microbial Technology*, **26**, 131–136.

Lilly, M., Lambrechts, M. G., and Pretorius, I. S. (2000). Effect of increased yeast alcohol acetyltransferase activity on flavor profiles of wine and distillates. *Applied and Enviromental Microbiology*, **66**, 744–753.

Olempska-Beer, Z. S., Merker, R. I., Ditto, M. D., and DiNovi, M. J. (2006). Food-processing enzymes from recombinant microorganisms—A review. *Regulatory Toxicology and Pharmacology*, **45**, 144–158.

Pandey, A. and Ramachandran, S. (2005). *Enzyme technology*. New Delhi: Asiatech Publishers, 1–10.

Rehman, S., Paterson, A., and Piggott, J. R. (2006). Flavour in sourdough breads: A review. *Trends in Food Science and Technology*, **17**, 557–566.

Rossi, S. C., Vandenberghe, L. P. S., Pereira, B. M. P., Gago, F. D., Rizzolo, J. A., Pandey, A., Soccol, C. R., and Medeiros, A. B. P. (2009). Improving fruity aroma production by fungi in SSF using citric pulp. *Food Research International*, **42**, 484–486.

Salah, R. B., Ghamghui, H., Miled, N., Mejdoub, H., and Gargouri, Y. (2007). Production of butyl acetate ester by lipase from Novel Strain of *Rhizopus oryzae*, *Journal of Biosciense and Bioengineering*, **103**, 368–372.

Soccol, C. R., Medeiros, A. B. P., Vandenberghe, L. P. S., and Woiciechowski, A. L. (2007). Flavor compounds produced by fungi, yeasts and bacteria. In Y. H. Hui. (Org.), *Flavor Technology* (pp. 179–191). Hoboken, NJ: John Wiley and Sons.

Soccol, C. R., Medeiros, A. B. P., Vandenberghe, L. P. S., Soares, M., and Pandey, A. (2008). Production of aroma compounds. *Current Developments in Solid-State Fermentation* (pp. 357–372). New Delhi: Springer Asiatech Publishers.

Soccol, C. R., Vandenberghe, L. P. S, Woiciechowski, A. L., and Babitha, S. (2005). Applicatios of industrial enzymes. *Enzyme Technology* (p. 742). New Delhi: Asiatech Publishers.

Thakore, Y. (2008). *Report ID:BIO030E*, Published: January 2008, Analyst: Business Communications Company, Inc.

Torres, S., Baigorí, M. D., Swathy, S. L., Pandey, A., and Castro, G. R. (2009). Enzymatic synthesis of banana flavour (isoamyl acetate) by *Bacillus licheniformis* S-86 esterase. *Food Research International*, **42**, 454–460.

Van Laere, S. D. M., Saerens, S. M. G., Verstrepen, K. J., Van Dijck, P., Thevelein, J. M., and Delvaux, F. R. (2008). Flavour formation in fungi: Characterisation of KlAtf, the *Kluyveromyces lactis* orthologue of the *Saccharomyces cerevisiae* alcohol acetyltransferases *Atf1* and *Atf2*. *Applied Microbiology Biotechnology*, **78**, 783–792.

Vandamme, E. J. and Soetaert, W. (2002). Bioflavours and fragrances via fermentation.

Xu, Y., Wang, D., Mu, X. Q., Zhao, G. A., and Zhang, K. C. (2002). Biosynthesis of ethyl esters of short-chain fatty acids using whole-cell lipase from *Rhizopus chinesis* CCTCC M201021 in non-aqueous phase. *Journal of Molecular Catalylis Enzymatic*, **18**, 29–37.

2

Ahmed, J. and Ramaswamy, H. S. (2006). High pressure processing of fruits and vegetables. *Stewart Postharvest Review*, **1**, 1–10.

Chen, Z, Liu, Y. M., Yang, S., Song, B. A., Xu, G. F., Bhadury, P. S., Jin, L, H., Hu, D. Y., Liu, F., Xue, W., and Zhou, X. (2008). Studies on the chemical constituents and anticancer activity of *Saxifraga stolonifera* (L) Meeb. *Bioorganic & Medicinal Chemistry*, **16**, 1337–1344.

Corrales, M., Toepfl, S., Butz, P., Knorr, D., and Tauscher, B. (2008). Extraction of anthocyanins from grape by-products assisted by ultrasonic, high hydrostatic pressure or pulsed electric fields: *A comparison, Innovative Food Science and Emerging Technology*, **9**, 85–91.

Dornenburg, H. and Knoor, D. (1993). Cellular permeability of cutured plant tissues by high electric field pulses or ultra high pressure for the recovery of secondary metabolites. *Food Biotechnology*, **7**, 35–48.

Duan, X., Wu, G., and Jiang, Y. (2007a). Evaluation of antioxidant properties of phenolics from litchi fruit in relation to pericarp browning prevention. *Molecules*, **12**, 759–771.

Duan, X., Jiang, Y., Su, X., Zhang, Z., and Shi, J. (2007b). Antioxidant properties of anthocyanins extracted from litchi (*Litchi sinensis* Sonn.) fruit pericarp tissues in relation to their role in the pericarp browning. *Food Chemistry*, **101**, 1365–1371.

Engelberth, A. S., Clausen, E. C., and Carrier, D. J. (2010). Comparing extraction method to recover ginseng saponins from American ginseng (I), followed by purification using fast centrifugal partition chrotomography with HPLC verification. *Separation and Purification Technology*, **72**, 1–6.

Guo, C., Yang, J., Wei, J., Li, Y., Xu, J., and Jiang, Y. (2003). Antioxidant activities of peel, pulp and seed fractions of common fruits as determined by FRAP assay. *Nutrition Research*, **23**, 1719–1726.

Hall, N. T., Smoot, J. M., Knight, R. J. Jr., and Nagy, S. (1980). Protein and amino acid compositions of ten tropical fruits by gas-liquid chromatography. *Journal of Agriculture and Food Chemistry*, **28**, 1217–1221.

Halliwell, B. (2007). Dietary polyphenols: Good, bad, or indifferent for your health? *Cardiovascular Research*, **73**, 341–347.

Jiang, Y., Wang, Y., Song, L., Liu, H., Lichter, A., Kerdchoechuen, O., Joyce, C., and Shi, S. (2006). Production and postharvest characteristics and technology of litchi fruit: an overview. *Australian Journal of Experimental Agriculture*, **46**, 1541–1556.

Khan, I. U., Asghar, M. N., Iqbal, S., and Bokhari, T. H. (2009). Racical savenging and antioxidant potential of aqueous and organic extracts of aerial parts of *Litchi chinensis* sonn. *Asian Journal of Chemistry*, **21**, 5073–5084.

Knorr, D., Ade-Omowaye, B. I. O., and Heinz, V. (2002). Nutritional improvement of plant foods by non-thermal processing. *Proceedings of Nutrition Society*, **61**, 311–318.

Kong, F. L., Zhang, M. W., Kuang, R. B., Yu, S. J., Chi, J. W., and Wei, Z. C. (2010). Antioxidant activities of different fractions 1 of polysaccharide purified from pulp tissue of litchi (*Litchi chinensis* Sonn.). *Carbohydrate Polymers*, Doi:10.1016/j.carbpol.2010.03.021.

Lee, H.S. and Wicker, L. (1990). Anthocyanin pigments in the skin of lychee fruit. *Journal of Food Science*, **56**, 466–468, 483.

Li, J. G. (2008). *In "The Litchi"*, China Agric. Press, Beijing, China (in Chinese).

Li, J. and Jiang, Y. (2007). Litchi flavonoids: isolation, identification and biological activity. *Molecules*, **12**, 745–758.

Liu, L., Xie, B., Cao, S., Yang, E., Xu, X., and Guo, S. (2007). A-type procyanidins from *Litchi chinensis* pericarp with antioxidant activity. *Food Chemistry*, **105**, 1446–1451.

Lou, Z., Wang, H., Zhang, M., and Wang, Z. (2010). Improved extraction of oil from chichea under ultrasound in a dynamic syste. *Journal of Food Engineering*, **98**, 13–19.

Mahattanatawee, K., Manthey, J. A., Luzio, G., Talcott, S. T., Goodner, K., and Baldwin, E. A. (2006). Total antioxidant activity and fiber content of select Florida-grown tropical fruits. *Journal of Agricultur and Food Chemistry*, **54**, 735–7363.

Mason, T. J. and Zhao, Y. (1994). Enhancement of ultrasonic cavitation yield by multi-frequency sonication. *Ultrasonic Sonochemistry*, **29**, 567–582.

Prasad, U. S. and Jha, O. P. (1978). Changes in pigmentation patterns during litchi ripening: flavonoid,. *The Plant Biochemical Journal*, **5**, 44–49.

Prasad, N, K., Yang, B., Shi, S., Yi, C., Zhao, M., Xue, S., and Jiang, Y. (2010) Enhanced antioxidant and antityrosinase activities of longan fruit pericarp by ultra-high pressure-assisted extraction processing. *Journal of Pharmaceutical and Biomedical Analysis*, **51**, 471–477.

Prasad, N. K., Yang, B., Zhao, M., Wang, B., Chen, F., and Jiang, Y. (2009 a). Effects of high pressure treatment on the extraction yield, phenolic content and antioxidant activity of litchi (*Litchi chinensis* Sonn.) fruit pericarp. *International Journal of Food Science and Technology*, **44**, 960–966.

Prasad, N, K., Yang, B., Zhao, M., Ruenroengklin, N., and Jiang, Y. (2009 b). Application of ultrasonication or high pressure extraction of flavonoids from litchi fruit pericarp. *Journal of Food Process Engineering*, **32**, 828–843.

Prasad, N. K., Yang, B., Yang, S., Chen, Y., Zhao, M., Ashraf, M., and Jiang, Y. (2009c). Identification of phenolic compounds and appraisal of antioxidant actityrosinase activity of litchi (*Litchi sinensis* Sonn.) seeds. *Food Chemistry*, **116**, 1–7.

Rao G, Shen, G., and Xu, G. (2009). Ultrasonic assisted extraction of coenzyme Q10 from litchi (*Litchi chinensis* sonn.) pericarp using response surface methodology. *Journal of Food Process Engineering*, doi: 10.1111/j.1745-4530.2009.00420.x

Roux E. L., Doco, T., Sarni-Manchado, P., Lozano, Y., and Cheynier, V. (1998). A-type proanthocyanidins from pericarp of *Litchi sinensis*. *Phytochemistry*, **48**, 1251–1258.

Ruenroengklin, N., Zhong, J., Duan, X., Yang, B., Li, J., and Jiang, Y. (2008). Effects of various temperatures and pH values on the extraction yield of phenolics from litchi fruit pericarp tissue and the antioxidant activity of the extracted anthocyanins. *International Journal of Molecular Sciences*, **9**, 1333–1341.

Sarni-Manchado, P., Roux, E. L., Guerneve, C. L., Lozano Y., and Cheynier, V. (2000). Phenolic composition of litchi fruit pericarp. *Journal of Agriculture and Food Chemistry*, **48**, 5995–6002.

Sivakumar, D., Naude, Y., Rohwer, E., and Korsten, L. (2008). Volatile compounds, quality attributes, mineral composition and pericarp structure of South African litchi export cultivars Maurities and MacLean's Red. *Journal of the Science of Food and Agriculture*, **88**, 1074–1081.

Sun, J., Jiang, Y., Shi, S., Wei, X., Xue, S. H., Shi, J., and Yi, C. (2010). Antioxidant activities and contents of polyphenol oxidase substrates from pericarp tissues of litchi fruit. *Food Chemistry*, **119**, 753–757

US FDA. (2000). Kinetics of microbial inactivation for alternative food processing technologies-high pressure processing. Retrieved from http://vm.cfsan.fda.gov/-comm/ifthpp.html

Wall, M. M. (2006). Ascorbic acid and mineral composition of longan (*Dimocarpus longan*), lychee (*Litchi chinensis*) and rambutan (*Nephelium lappaceum*) cultivars grown in Hawaii. *Journal of Food Composition and Analysis*, **19**, 655–663.

Wang, X., Wei, Y., Yuan, S., Liu, G., Zhang, Y. L. J., and Wang, W. (2006a). Potential anticancer activity against of litchi pericarp water extract against hepatocellular carcinoma using *in-vitro* and *in-vivo*. *Cancer Letters*, **239**, 144–150. Retrieved from http://vm.cfsan.fda.gov/-comm/ifthpp.html

Wang, X., Yuan, S., Wang, J., Lin, P., Liu, G., Lu, Y., Zhang, J., Wang, W., and Wei, Y. (2006b), Anticancer activity of litchi fruit pericarp extract against human breast cancer *in vitro* and *in vivo*. *Toxicology and Applied Pharmacology*, **215**, 168–178.

Xiao, C. (2003). *Chemistry of Chinese Medicine*, Shanghai Science and Technology Press, Shanghai, China (In Chinese).

Yang, B., Wang, J., Zhao, M., Liu, Y., Wang, W., and Jiang, Y. (2006). Identification of polysaccharides from pericarp tissues of litchi (*Litchi chinensis* Sonn.) fruit in relation to their antioxidant activities. *Carbohydrate Research,* **341**, 634–638.

Yang, B. Zhao, M. Shi, J. Yang, N., and Jiang, Y. (2008). Effect of ultrasonic treatment on the recovery and DPPH radical scavenging activity of polysaccharides from longan fruit pericarp. *Food Chemistry*, **106**, 685–690.

Zhang, S., Junjie, Z., and Changzhen, W. (2004b). Novel high pressure extraction technology. *International Journal of Pharmaceutics*, **78**, 471–474.

Zhang, D. L., Quantick, P. C., and Grigor, J. M. (2000). Changes in phenolic compounds in litchi (*Litchi chinensis* Sonn.) fruit during posthravest storage. *Postharvest Biology and Technology*, **19**, 165–172.

Zhang, Z, Xuequn, P., Yang, C., Ji, Z., and Jiang, Y. (2004a). Purification and structural analysis of anthocyanins from litchi pericarp. *Food Chemistry*, **84**, 601–604.

Zhao, M., Yamg, B., Wang, J., Li, B., and Jiang, Y. (2006). Identification of the major flavonoids from pericarp tissues of lychee fruit in relation to their antioxidant activities. *Food Chemistry*, **98**, 539–544.

Zhao, M., Yang, B., Wang, J., Liu, Y., Yu, L., and Jiang, Y. (2007). Immunomodulatory and anticancer activities of flavonoids extracted from litchi (*Litchi chinensis* Sonn.) pericarp. *International Immunopharmacology*, **7**, 162–166.

3

Aisa, H. A., Zhen, C., Yili, A., Bahang, R., Rakhmanberdyeva, R. K., and Sagdullaev, B. T. (2006). Polysaccharides from two *Cicer* species cultivated in China. *Chemistry of Natural Compounds*, **42**, 349–350.

Cardoso, S. M., Coimbra, M. A., and Lopes da Silva, J. A. (2003). Calcium-mediated gelation of an olive pomace pectin extract. *Carbohydrate Polymers*, **52**, 125–133.

Carvajal-Millán, E., Guigliarelli, B., Belle, V., Rouau, X., and Micard, V. (2005). Storage stability of arabinoxylan gels. *Carbohydrate Polymers*, **59**, 181–188.

Carvajal-Millán, E., Rascón-Chu, A., Márquez-Escalante, J., Ponce de León, N., Micard, V., and Gardea, A. (2007). Maize bran gum: Extraction, characterization and functional properties. *Carbohydrate Polymers*, **6**, 280–285.

Christodoulou, V., Bampidis, V. A., Hucko, B., Ploumi, K., Iliadis, C., Robinson, P. H., and Mudrik, Z. (2005). Nutritional value of chickpeas in rations of lactating ewes and growing lambs. *Animal Feed Science Technology*, **118**, 229–241.

Doublier, J. L. and Cuvelier, G. (1996). Gums and hydrocolloids: Functional aspects. In A. C. Eliasson (Ed.), *Carbohydrates in Food* (pp. 283–318*)*. New York: Marcel Dekker, Inc.

Gnanasambandan, R. and Proctor, A. (2000). Determination of pectin degree of esterification by diffuse reflectance Fourier transform infrared spectroscopy. *Food Chemistry*, **68**, 327–332.

Kraemer, E. O. (1938). Molecular weight of celluloses and cellulose derivates. *Industrial and Engineering Chemistry*, **30**, 1200–1203.

Maheri-Sis, N., Chamani, M., Sadeghi, A. A., Mirza-Aghazadeh, A., and Aghajanzadeh-Golshani, A. (2008). Nutritional evaluation of kabuli and desi type chickpeas (*Cicer arietinum* L.) for nutriments using *in vitro* gas production technique. *African Journal of Biotechnology*, **7**, 2946–2951.

May, C. D. (1990). Industrial pectins: Sources, production and applications. *Carbohydrate Polymers*, **12**, 79–99.

Mead, D. J. and Fouss, R. M. (1942). Viscosities of solutions of polyvinyl chloride. *Journal of the American Chemical Society*, **64**, 277–282.

Ramos-Chavira, N., Carvajal-Millan, E., Marquez-Escalante, J., Santana-Rodriguez, V., Rascon-Chu, A., and Salmerón-Zamora, J. (2009). Characterization and functional properties of an oat gum extracted from a drought harvested *A. sativa*. *Journal of Food Science and Biotechnology*, **18**, 900–903.

Rascón-Chu, A., Martínez-López, A. L., Carvajal-Millán, E., Ponce de León-Renova, N., Márquez-Escalante, J., and Romo-Chacón, A. (2009). Pectin from low quality 'Golden Delicious' apples: Composition and gelling capability. *Food Chemistry*, **116**, 101–103.

Ross-Murphy, S. B. (1984). Rheological methods. In H. W. S. Chan (Ed.), *Biophysical Methods in Food Research* (pp. 138–199). Oxford: Blackwell.

Ström, A., Ribelles, P., Lundin, L., Norton, I., Morris, E. R., and Williams, A. K. (2007). Influence of pectin fine structure on the mechanical properties of calcium-pectin and acid-pectin gels. *Biomacromolecules*, **8**, 2668–2674.

Temelli, F. (1997). Extraction and functional properties of barley β-glucan as affected by temperature and pH. *Journal of Food Science*, **62**, 1194–1198.

Willats, W. G. T., Paul Knox, J., and Mikkelsen, J. D. (2006). Pectin: New insights into an old polymer are starting to gel. *Trends in Food Science and Technology*, **17**, 97–104.

Yapo, B. M. and Koffi, K. L. (2006). Yellow passion fruit rinds a potential source of low-methoxyl pectin. *Journal of Agricultural and Food Chemistry*, **54**, 2738–2744.

4

Afidah, A. R., Rocca, E., Steinmetz, J. M., Jain, K. M., Sani, I., and Hasnah, O. (2008). Antioxidant activities of mangrove *Rhizophora apiculata* bark extracts. *Food Chemistry*, **107**, 200–207.

Aline, M., Charles, E. L., Marco, R., Jeanne, M., and Odile, G. N. (2005). Determination of the total phenolic, flavonoid, and proline contents in *Burkina Fasan* honey, as well as their radical scavenging activity. *Food Chemistry*, **91**, 571–577.

Amin, I., Zamaliah, M. M., and Chin, W. F. (2004). Total antioxidant activity and phenolic content in selected vegetables. *Food Chemistry*, **87**, 581–586.

Block, G. (1992). A role for antioxidants in reducing cancer risk. *Nutrition Reviews*, **50**, 207–213.

Bors, W. and Saran, M. (1987). Radical scavenging by flavonoid antioxidants. *Free Radical Research Communication*, **2**(4–6), 289–294.

Deachathai, S., Mahabusarakam, W., Phongpaichit, S., and Taylor, W. C. (2005). Phenolic compounds from the fruit of *Garcinia dulcis*. *Phytochemicals*, **66**, 2368–2375.

Deachathai, S., Mahabusarakam, W., Phongpaichit, S., Taylor, W. C., Zhang, Y. J., and Yang, C. R. (2006). Phenolic compounds from the flowers of *Garcinia dulcis*. *Phytochemicals*, **67**, 464–469.

Emmons, C. L. and Peterson, D. M. (2001). Antioxidant activity and phenolic content of oat as affected by cultivar and location. *Crop Science*, **41**, 1676–1681.

Gabor, M. (1986). Anti-inflammatory and antiallergic properties of flavonoids. *Progress in Clinical and Biological Research*, **213**, 471–480.

Giusti, M. M. and Wrolstad, R. E. (2001). Anthocyanins: Characterization and measurement of anthocyanins by UV–Visible spectroscopy. In R. E. Wrolstad (Ed.), *Current protocols in food analytical chemistry*, New York: John Wiley & Sons Unit F1 2:1–13.

Gopalakrishnan, C., Banumathi, D., and Suresh, S. K. (1997). Effect of mangostin, a xanthone from *Garcinia mangostana* Linn. in immunopathological and inflammatory reactions. *Indian Journal of Experimental Biology*, **19**, 843–846.

Grassman, J., Hippeli, S., and Elstner, E. F. (2002). Plant's defence and its benefit for animal and medicine: Role of phenolics and terpenoids in avoiding oxygen stress. *Plant Physiology and Biochemistry*, **40**, 471–478.

Husain, S. R., Cillard, J., and Cillard, P. (1987). Hydroxyl radical scavenging activity of flavonoids. *Phytochemicals*, **26**, 2489–2497.

Jayaprakasha, G. K., Negi, P. S., and Jena, B. S. (2006). Antioxidative and antimutagenic activities of the extracts from the rinds of *Garcinia pedunculata*. *Innovative Food Science & Emerging Technology*, **7**, 246–250.

Jiangrong, Li. and Yueming, J. (2007). Litchi Flavonoids: Isolation, identification, and biological activity. *International Journal of Molecular Sciences*, **12**, 745–758.

Kähkönen, M. P., Hopia, A. I., Vuorela, H. J., Rauha, J. P., Pihlaja, K., Kujala, T. S., and Heinonen, M. (1999). Antioxidant activity of plant extracts containing phenolic compounds. *Journal of Agricultural and Food Chemistry*, **47**, 3954–3962.

Kalyarat, K. and Kaew, K. (2006). Antioxidant activity, phenolic compound contents and antimutagenic activity of some water extract of herbs. *Thai Journal of Pharmaceutical Sciences*, **30**, 28–35.

Konczak, I. and Zhang, W. (2004). Anthocyanins-more than nature's colours. *Journal of Biomedical and Biotechnology*, **5**, 239–240.

Kong, J. M., Chia, L. S., Goh, N. K., Chia, T. F., and Brouillard, R. (2003). Analysis and biological activities of anthocyanins, *Phytochemicals*, **64**(5), 923–933.

Kosin, J., Ruangrungsi, N., Ito, C., and Furukawa, H. (1998). A xanthone from *Garcinia atroviridis*. *Phytochemicals*, **47**, 1167–1168.

Kovacs, E. M. R., Westerterp-Plantenga, M. S., and Saris, W. H. M. (2001). The effects of 2-week ingestion of ()-hydroxycitrate and ()-hydroxycitrate combined with medium-chain triglycerides on satiety, fat oxidation, energy expenditure, and body weight. *International Journal of Obesity and Related Metabolic Disorders*, **25**, 1087–1094.

Lewis, Y. S. and Neelakantan, S. (1965). ()-Hydroxycitric acid–the principle acid in the fruits of *Garcinia cambogia Desr*. *Phytochemicals*, **4**, 619–625.

Mackeen, N. M., Ali, A. M., Lajis, N. H., Kawazu, K., Hassan, Z., Amran, M., Habsah, M., Mooi, L. Y., and Mohamed, S. M. (2000). Antimicrobial, antioxidant, antitumour-promoting and cytotoxic activities of different plant part extracts of *Garcinia atroviridis griff* ex Tanders. *Journal of Ethnopharmacology*, **72**, 395–402.

Mahabusarakum, W., Phongpaichit, S., Jansakul, C., and Wiriyachitra, P. (1983). Screening of antibacterial activity of chemicals from *Garcinia mangostana* Songklanakarin. *Journal of Science and Technology*, **5**, 337–339.

Middleton, E. J., Kandaswami, C., and Theoharides, T. C. (2000). The effects of plant flavonoids on mammalian cells: implication for inflammation, heart diseases, and cancer. *Pharmacological Reviews*, **52**, 673–651.

Namiki, M. (1990). Antioxidant/antimutagens in food. *CRC Critical Review Food Science and Nutrition*, **29**, 273–300.

Nor Hadiani, I. and Khozirah, S. (2008). Beyond medicinal plants, reality, and challenges in antidiebetic research. University Publication Centre (UPENA) UiTM, 153–157.

Nour-Eddine, E. S., Souhila, G., and Paul, H. D. (2007). Flavonoids: Hesynthesis, reactivity, characterization, and free radical scavenging activity. *International Journal of Molecular Sciences*, **12**, 2228–2258.

Orsolya, F., Judit, J., and Károly, H. (2004). Antioxidant activity relationships of flavonoids compounds. *International Journal of Molecular Sciences*, **9**, 1079–1088.

Papetti, G. A., Massolini, G., and Daglia, M. (1998). Anti- and prooxidant activity of water soluble components of some common diet vegetables and the effect of thermal treatment. *Journal of Agriculture and Food Chemistry*, **46**, 4118–4122.

Permana, D., Lajis, N. H., Mackeen, M. M., Ali, A. M., Aimi, N., Kitajima, M., and Takayama, H.

(2001). Isolation and bioactivities of constituents of the roots of *Garcinia atroviridis*. *Journal of Natural Products*, **64**, 976–979.

Rafael, C. D., Magda, N. L., and Nadia, R. B. (2008). Quantification of phenolic of Pterodon emarginatus vogel seeds. *International Journal of Molecular Sciences*, **9**, 606–614.

Sakagami, Y., Iinuma, M., Piyasena, K. G. N. P. and Dharmaratne, H. R. W. (2005). Antibacterial activity of a-mangostin against vancomycin resistant Enterococci (VRE) and synergism with antibiotics. *Phytomedicine*, **12**, 203–208.

Singh, R. B., Ghosh, S., and Niaz, M. A., et al. (1995). Dietary intake, plasma levels of antioxidant vitamins, and oxidative stress in relation to coronary artery disease in elderly subjects. *American Journal of Cardiology*, **76**, 1233–1238.

Sri Nurestri, A. M., Norhanom, A. W., Hashim, Y., Sim, K. S., Hong, S. L., Lee, G. S., and Syarifah, N. S. A. R. (2008). Cytotoxic activity of *Pereskia bleo* (*Cactaceae*) against selected human cell lines. *International Journal of Cancer Research*, **4**, 20–27.

Stintzing, F. C. and Carle, R. (2004). Functional properties of anthocyanins and betalains in plants, foods, and in human nutrition. *Trends in Food Science and Technology*, **15**(1), 19–38.

Sullivan, A. C., Triscari, J., Hamilton, J. G., Miller, O. N., and Wheatley, V. R. (1974). *Lipids*, **9**, 121–128.

Sze-Tao, K. W. C., Schrimpf, J. E., Teuber, S. S., Roux, K. H., and Sathe, S. K. (2001). Effects of processing and storage on walnut (*Juglans regia L.*) tannins. *Journal of Science and Food Agricultural*, **81**(13), 1215–1222.

Williams, P., Ongsakul, M., Proudfoot, J., Croft, K., and Bellin, L. (1995). Mangostin inhibits the oxidative modification of human low density lipoprotein. *Free Radical Research*, **23**, 175–184.

Zadernowski, R., Naczk, M., and Nesterowicz, J. (2005). Phenolic acid profiles in some small berries. *Journal of Agricultural and Food Chemistry*, **53**(6), 2118–2124.

5

AePark, S., Choi, M., Cho, S., Seo, J., Jung, U., Kim, M., Sung, M., Park, Y., and Lee, M. (2006). Genistein and daidzein modulate hepatic glucose and lipid regulating enzyme activities in C57BL/KsJ-db/db mice. *Life Sciences*, **79**,1207–1213.

Agouni, A., Lagrue-Lak-Hal, A., Mostefai, H. A., Tesse, A., Mulder, P., Rouet, P., Desmoulin, F., Heymes, C., MartÃnez, M. C., and Andriantsitohaina, R. (2009). Red wine polyphenols prevent metabolic and cardiovascular alterations associated with obesity in zucker fatty rats (Fa/Fa). *PLoS ONE*, **4**, e55–57.

Aherne, S. A. and O'Brien, N. M. (2002). Dietary flavonols: Chemistry, food content, and metabolism. *Nutrition*, **18**, 75–81.

Al-Awwadi, N., Araiz, A. C., Bornet, A., Delbosc, S., Cristol, J., Linck, N., Azay, J., Teissedre, P., and Cros, G. (2005). Extracts enriched in different polyphenolic families normalize increased cardiac NADPH oxidase expression while having differential effects on insulin resistance, hypertension, and cardiac hypertrophy in high-fructose-fed rats. *Journal of Agricultural and Food Chemistry*, **53**, 1–51.

Al-Awwadi, N., Azay, J., Poucheret, P., Cassanas, G., Krosniak, M., Auger, C. A., Gasc, F., Rouanet, J., Cros, G., and Teissedre, P. L. (2004). Antidiabetic activity of red wine polyphenolic extract, ethanol, or both in streptozotocin-treated rats. *Journal of Agricultural and Food Chemistry*, **52**, 1008–1016.

Allen, R. R., Carson, L., Kwik-Uribe, C., Evans, E. M., and Erdman, J. W. Jr. (2008). Daily consumption of a dark chocolate containing flavanols and added sterol esters affects cardiovascular risk factors in a normotensive population with elevated cholesterol. *Journal of. Nutrition*, **138**, 725–731.

Amin, A. and Buratovich, M. (2007). The anticancer charm of flavonoids: A cup-of-tea will do! *Recent Patents on Anticancer Drug Discovery*, **2**, 109–117.

Aron, P. M. and Kennedy, J. A. (2008). Flavan-3-ols: Nature, occurrence and biological activity. *Molecular Nutrition & Food Research*, **52**, 79–104.

Arts, I. C. and Hollman, P. C. (2005). Polyphenols and disease risk in epidemiologic studies, *American Journal of Clinical Nutrition*, **81**, 317S–325S.

Avellone, G., Di Garbo, V., Campisi, D., De Simone, R., Raneli, G., Scaglione, R., and Licata, G. (2005). Effects of moderate sicilian red wine consumption on inflammatory biomarkers of atherosclerosis. *Europe Journal of Clinical Nutrition*, **60**, 41–47.

Aviram, M., Dornfeld, L., Rosenblat, M., Volkova, N., Kaplan, M., Coleman, R., Hayek, T., Presser, D., and Fuhrman, B. (2000). Pomegranate juice consumption reduces oxidative stress, atherogenic modifications to LDL, and platelet aggregation: Studies in humans and in atherosclerotic apolipoprotein E-deficient mice. *American Journal of Clinical Nutrition*, **71**, 1062–1076.

Baba, S., Osakabe, N., Kato, Y., Natsume, M., Yasuda, A., Kido, T., Fukuda, K., Muto, Y., and Kondo, K. (2007). Continuous intake of polyphenolic compounds containing cocoa powder reduces LDL oxidative susceptibility and has beneficial effects on plasma HDL-cholesterol concentrations in humans. *American Journal of Clinical Nutrition*, **85**, 709–717.

Baiges, I., Palmfeldt, J., Blade, C., Gregersen, N., and Arola, L. (2010). Lipogenesis is decreased by grape seed proanthocyanidins according to liver proteomics of rats fed a high fat diet. *Molecular & Cellular Proteomics*, **9**(7), 1499–1513.

Bajaj, M. and DeFronzo, R. (2003). Metabolic and molecular basis of insulin resistance. *Journal of Nuclear Cardiology*, **10**, 311–323.

Banini, A. E., Boyd, L. C., Allen, J. C., Allen, H. G., and Sauls, D. L. (2006). Muscadine grape products intake, diet and blood constituents of non-diabetic and type 2 diabetic subjects. *Nutrition*, **22**, 1137–1145.

Bhagwat, S., Holden, J., Haytowitz, D., Gebhardt, S., Eldridge, A., Dwyer, J., and Peterson, J. (2010). USDA database for the flavonoid content of selected foods.

Bladé, C., Arola, L., and Salvadó, M. (2010). Hypolipidemic effects of proanthocyanidins and their underlying biochemical and molecular mechanisms. *Molecular Nutrition & Food Research*, **54**, 37–59.

Blay, M., Pinent, M., Baiges, I., Ardévol, I., Salvadó, M., Bladé, C., and Arola, L. (2003). Grape seed proanthocyanidins increase glucose uptake in muscular cell line L6E9. In A. Lonvaud-Funel, G. deRevel, and P. Darriet (Eds.), *Oenologie '03. 7th International Symposium of Oenology* (pp. 669–673). Bordeaux: Editions Tec&Doc.

Bose, M., Lambert, J. D., Ju, J., Reuhl, K. R., Shapses, S. A., and Yang, C. S. (2008). The major green tea polyphenol, ()-epigallocatechin-3-gallate, inhibits obesity, metabolic syndrome, and fatty liver disease in high-fat-fed mice. *Journal of Nutrition*, **138**, 1677–1683.

Brand-Miller, J., Holt, S. H. A,. de Jong, V., and Petocz, P. (2003). Cocoa powder increases postprandial insulinemia in lean young adults. *Journal of Nutrition*, **133**, 3149–3152.

Cao, H., Hininger-Favier, I., Kelly, M. A., Benaraba, R., Dawson, H. D., Coves, S., Roussel, A. M., and Anderson, R. A. (2007). Green tea polyphenol extract regulates the expression of genes involved in glucose uptake and insulin signaling in rats fed a high fructose diet. *Journal of Agricultural and Food Chemistry*, **55**, 6372–6378.

Castell, A., Cedó, L., Pallarès, V., Blay, M., Pinent, M., García-Vallvé, S., Arola, L., and Ardévol, A. (2009). GSPE modify insulin synthesis and secretion in b-cell (Abstract). *Journal of Diabetes*, **1**, suppl. 1, A2–71.

Castilla, P., Echarri, R., Davalos, A., Cerrato, F., Ortega, H., Teruel, J. L., Lucas, M. F., Gomez-Coronado, D., Ortuno, J., and Lasuncion, M. A. (2006). Concentrated red grape juice exerts antioxidant, hypolipidemic, and antiinflammatory effects in both hemodialysis patients and healthy subjects. *American Journal of Clinical Nutrition*, **84**, 252–262.

Chou, E. J., Keevil, J. G., Aeschlimann, S., Wiebe, D. A., Folts, J. D., and Stein, J. H. (2001). Effect of ingestion of purple grape juice on endothelial function in patients with coronary heart disease. *American Journal of Cardiology*, **88**, 553–555.

Cordain, L., Melby, C. L., Hamamoto, A. E., O'Neill, D. S., Cornier, M., Barakat, H. A., Israel, R. G., and Hill, J. O. (2000). Influence of moderate chronic wine consumption on insulin sensitivity and other correlates of syndrome X in moderately obese women. *Metabolism Clinical Experimental*, **49**, 1473–1478.

Corder, R., Mullen, W., Khan, N. Q., Marks, S. C., Wood, E. G., Carrier, M. J., and Crozier, A.

(2006). Oenology: Red wine procyanidins and vascular health, *Nature*, **444**, 5–66.

Decorde, K., Teissedre, P. L., Sutra, T., Ventura, E., Cristol, J. P., and Rouanet, J. M. (2009). Chardonnay grape seed procyanidin extract supplementation prevents high-fat diet-induced obesity in hamsters by improving adipokine imbalance and oxidative stress markers. *Molecular Nutrition & Food Research*, **53**, 659–666.

Diepvens, K., Kovacs, E. M. R., Vogels, N., and Westerterp-Plantenga, M. S. (2006). Metabolic effects of green tea and of phases of weight loss, *Physiology & Behavior*, **87**, 185–191.

El-Alfy, A. T., Ahmed, A. A., and Fatani, A. J. (2005). Protective effect of red grape seeds proanthocyanidins against induction of diabetes by alloxan in rats. *Pharmacological Research*, **52**, 264–270.

Faria, A., Calhau, C., deFreitas, V., and Mateus, N. (2006). Procyanidins as antioxidants and tumor cell growth modulators. *Journal of Agricultural and Food Chemistry*, **54**, 2392–2397.

Fernandez-Larrea, J., Montagut, G., Bladé, C., Salvadó, M. J., Blay, M., Pujadas, G., Arola, L., and Ardévol, A. (2007). GSPE has the same effects as insulin on the mRNA levels of the main genes of glucose disposal in the liver of STZ-diabetic animals (Abstract). *Diabetes and Vascular Disease Research*, Mar;4 Suppl. 1, S1–86.

Grassi, D., Desideri, G., Necozione, S., Lippi, C., Casale, R., Properzi, G., Blumberg, J. B., and Ferri, C. (2008). Blood pressure is reduced and insulin sensitivity increased in glucose-intolerant, hypertensive subjects after 15 days of consuming high-polyphenol dark chocolate. *Journal of Nutrition*, **138**, 1671–1676.

Grassi, D., Lippi, C., Necozione, S., Desideri, G., and Ferri, C. (2005). Short-term administration of dark chocolate is followed by a significant increase in insulin sensitivity and a decrease in blood pressure in healthy persons. *American Journal of Clinical Nutrition*, **81**, 611–614.

Grassi, D., Necozione, S., Lippi, C., Croce, G., Valeri, L., Pasqualetti, P., Desideri, G., Blumberg, J. B., and Ferri, C. (2005). Cocoa reduces blood pressure and insulin resistance and improves endothelium-dependent vasodilation in hypertensives. *Hypertension*, **46**, 398–405.

Gu, L., Kelm, M., Hammerstone, J., Beecher, G., Holden, J., Haytowitz, D., Gebhardt, S., and Prior R. (2004). Concentrations of proanthocyanidins in common foods and estimations of normal consumption. *Journal of Nutrition*, **134**, 613–617.

Hossain, S. J., Kato, H., Aoshima, H., Yokoyama, T., Yamada, M., and Hara, Y. (2002). Polyphenol-induced inhibition of the response of Na+/Glucose cotransporter expressed in xenopus oocytes. *Journal of Agricultural and Food Chemistry*, **50**, 5215–5219.

Ichinose, K., Kawasaki, E., and Eguchi, K. (2007). Recent advancement of understanding pathogenesis of type 1 diabetes and potential relevance to diabetic nephropathy. *American Journal of Nephrology*, **27**, 554–564.

Jiménez, J. P., Serrano, J., Tabernero, M., Arranz, S., Díaz-Rubio, M. E., García-Diz, L., Goñi, I., and Saura-Calixto, F. (2008). Effects of grape antioxidant dietary fiber in cardiovascular disease risk factors. *Nutrition*, **24**, 646–653.

Jung, U., Lee, M., Park, Y., Kang, M., and Choi, M. (2006). Effect of citrus flavonoids on lipid metabolism and glucose-regulating enzyme mRNA levels in type-2 diabetic mice. *The International Journal of Biochemistry & Cell Biology*, **38**, 1134–1145.

Kar, P., Laight, D., Rooprai, H. K., Shaw, K. M., and Cummings, M. (2009). Effects of grape seed extract in type 2 diabetic subjects at high cardiovascular risk: A double blind randomized placebo controlled trial examining metabolic markers, vascular tone, inflammation, oxidative stress and insulin sensitivity. *Diabetic Medicine*, **26**, 526–531.

Karthikeyan, K., Bai, B. R. S., and Devaraj, S. N. (2007). Cardioprotective effect of grape seed proanthocyanidins on isoproterenol-induced myocardial injury in rats. *International Journal of Cardiology*, **115**, 326–333.

Kashyap, S. R. and DeFronzo, R. A. (2007). The insulin resistance syndrome: Physiological considerations. *Diabetes and Vascular Disease Research*, **4**, 9–13.

Kim, H., Park, H., Son, K., Chang, H., and Kang, S. (2008). Biochemical pharmacology of biflavonoids: Implications for anti-inflammatory action. *Archives of Pharmacal Research*, **31**, 265–273.

Kobayashi, Y., Suzuki, M., Satsu, H., Arai, S., Hara, Y., Suzuki, K., Miyamoto, Y., and Shimizu, M. (2000). Green tea polyphenols inhibit the sodium-dependent glucose transporter of intestinal epithelial cells by a competitive mechanism. *Journal of Agricultral and Food Chemistry*, **48**, 5618–5623.

Kottra, G. and Daniel, H. (2007). Flavonoid glycosides are not transported by the human Na+/glucose transporter when expressed in xenopus laevis oocytes, but effectively—inhibit electrogenic glucose uptake. *Journal of Pharmacology and Experimental Therapeutics*, **322**, 829–835.

Kwon, O., Eck, P., Chen, S., Corpe, C. P., Lee, J., Kruhlak, M., and Levine, M. (2007). Inhibition of the intestinal glucose transporter GLUT2 by flavonoids. *FASEB Journal*, **21**, 366–377.

Landrault, N., Poucheret, P., Azay, J., Krosniak, M., Gasc, F., Jenin, C., Cros, G., and Teissedre, P. L. (2003). Effect of a polyphenols-enriched chardonnay white wine in diabetic rats. *Journal of Agricultural and Food Chemistry*, **51**, 311–318.

Le, K. A. (2006). Metabolic effects of fructose. *Current Opinion in Clinical Nutrition & Metabolic Care*, **9**, 4–69.

Lee, Y., Cho, E., Tanaka, T., and Yokozawa, T. (2007). Inhibitory activities of proanthocyanidins from persimmon against oxidative stress and digestive enzymes related to diabetes. **53**, 287–292.

Lee, Y. A., Cho, E. J., and Yokozawa, T. (2008). Effects of proanthocyanidin preparations on hyperlipidemia and other biomarkers in mouse model of type 2 diabetes. *Journal of Agricultural and Food Chemistry*, **56**, 7781–7789.

Lee, Y., Kim, Y., Cho, E., and Yokozawa, T. (2007). Ameliorative effects of proanthocyanidin on oxidative stress and inflammation in streptozotocin-induced diabetic rats. *Journal of Agricultural and Food Chemistry*, **55**(23), 9395–9400.

Li, B. Y., Cheng, M., Gao, H. Q., Ma, Y. B., Xu, L., Li, X. H., Li, X. L., and You, B. A. (2008). Back-regulation of six oxidative stress proteins with grape seed proanthocyanidin extracts in rat diabetic nephropathy. *Journal of Cellular Biochemistry*, **104**, 668–679.

Li, W., Dai, R., Yu, Y., Li, L., Wu, C., Luan, W., Meng W., Zhang. X., and Deng, Y. (2007). Antihyperglycemic effect of cephalotaxus sinensis leaves and GLUT-4 translocation facilitating activity of its flavonoid constituents. *Biological & Pharmaceutical Bulletin*, **30**, 1123–1129.

Li, X. H., Xiao, Y. L., Gao, H. Q., Li, B. Y., Xu, L., Cheng, M., Jiang, B., and Ma, Y. B. (2009). Grape seed proanthocyanidins ameliorate diabetic nephropathy via modulation of levels of AGE, RAGE and CTGF. *Nephron Experimental Nephrology*, **111**, E31–41.

Liu, X., Wei, J., Tan, F., Zhou, S., Wurthwein, G., and Rohdewald, P. (2004). Antidiabetic effect of pycnogenol french maritime pine bark extract in patients with diabetes type II. *Life Sciences*, **75**, 2505–2513.

Liu, X., Zhou, H., and Rohdewald, P. (2004). French maritime pine bark extract pycnogenol dose-dependently lowers glucose in type 2 diabetic patients. *Diabetes Care*, **27**, 8–39.

Llópiz, N., Puiggròs, F., Céspedes, E., Arola, L., Ardévol, A., Bladé, C., and Salvadó, M. J. (2004). Antigenotoxic effect of grape seed procyanidin extract in fao cells submitted to oxidative stress. *Journal of Agricultural and Food Chemistry*, **52**, 10–83.

Loo, A. and Huang, D. (2007). Assay-guided fractionation study of alpha-amylase inhibitors from garcinia mangostana pericarp. **55**, 9805–9810.

Mantena, S. K., Baliga, M. S., and Katiyar, S. K. (2006). Grape seed proanthocyanidins induce apoptosis and inhibit metastasis of highly metastatic breast carcinoma cells. *Carcinogenesis*, **27**, 1682–1691.

Maritim, A. D. B., Sanders, R. A., and Watkins J. B. 3rd. (2003). Effects of pycnogenol treatment on oxidative stress in streptozotocin-induced diabetic rats. *Journal of Biochemical and Molecular Toxicology*, **17**, 193–199.

Montagut, G., Blade, C., Blay, M., Fernandez-Larrea, J., Pujadas, G., Salvado, M. J., Arola, L., Pinent, M., and Ardevol, A. (2009). Effects of a grapeseed procyanidin extract (GSPE) on insulin resistance. *Journal of Nutritional Biochemistry*, **21**, 961–967.

Montagut, G., Fernandez-Larrea, J., Romero, M., Esteve, M., Blade, C., Blay, M., Pujadas, G., Salvado, M. J., Arola, L., and Ardevol, A. (2007). Differential effects of grape-seed derived procyanidins on adipocyte differentiation markers in different *in vivo* situations. *Genes and Nutrition*, **2**, 101–103.

Montagut, G., Onnockx, S., Vaque, M., Blade, C., Blay, M., Fernandez-Larrea, J., Pujadas, G., Salvado, M. J., Arola, L., Pirson, I., Ardevol, A., and Pinent, M. (2009). Oligomers of grape-seed procyanidin extract activate the insulin receptor and key targets of the insulin signaling pathway differently from insulin. *Journal of Nutritional Biochemistry*, **21**(6), 476-481.

Muniyappa, R., Hall, G., Kolodziej, T. L., Karne, R. J., Crandon, S. K., and Quon, M. J. (2008). Cocoa consumption for 2 wk enhances insulin-mediated vasodilatation without improving blood pressure or insulin resistance in essential hypertension. *American Journal of Clinical Nutrition*, **88**, 1685–1696.

Osakabe, N., Yamagishi, M., Natsume, M., Yasuda, A., and Osawa, T. (2004). Ingestion of proanthocyanidins derived from cacao inhibits diabetes-induced cataract formation in rats. *Experimental Biology and Medicine*, **229**, 33–39.

Park, S. (2008). Dangnyohwan improves glucose utilization and reduces insulin resistance by increasing the adipocyte-specific GLUT4 expression in otsuka long-evans tokushima fatty rats. *Journal of Ethnopharmacology*, **115**, 473–482.

Petry, C. J., Ozanne, S. E., and Hales, C. N. (2001). Programming of intermediary metabolism. *Molecular Cellular Endocrinology*, **185**, 81–91.

Pinent, M., Blade, M. C., Salvado, M. J., Arola, L., Hackl, H., Quackenbush, J., Trajanoski, Z., and Ardevol, A. (2005). Grape-seed derived procyanidins interfere with adipogenesis of 3T3-L1 cells at the onset of differentiation. *International Journal of Obesity and Related Metabolic Disorders*, **29**, 934–941.

Pinent, M., Bladé, M., Salvad6, M., Blay, M., Pujadas, G., Fernàndez-Larrea, J., Arola, L., and Ardévol, A. (2005). Procyanidin effects on adipocyte-related pathologies. *Critical Reviews in Food Science and Nutrition*, **46**, 543–550.

Pinent, M., Blay M., Blade, M. C., Salvado, M. J., Arola, L., and Ardevol, A. (2004). Grape seed-derived procyanidins have an antihyperglycemic effect in streptozotocin-induced diabetic rats and insulinomimetic activity in insulin-sensitive cell lines. *Endocrinology*, **145**, 4985–4990.

Pinent, M., Castell, A., Baiges, I., Montagut, G., Arola, L., and Ardévol, A. (2008). Bioactivity of flavonoids on insulin-secreting cell. *Comprehensive Reviews in Food Science and Food Safety*, **7**, 299–308.

Potenza, M. A., Marasciulo, F. L., Tarquinio, M., Tiravanti, E., Colantuono, G., Federici, A., Kim, J., Quon, M. J., and Montagnani, M. (2007). EGCG, a green tea polyphenol, improves endothelial function and insulin sensitivity, reduces blood pressure, and protects against myocardial I/R injury in SHR. *American Journal of Physiology—Endocrinology and Metabolism*, **292**, E1378–1387.

Puiggros, F., Sala, E., Vaque, M., Ardevol, A., Blay, M., Fernandez-Larrea, J., Arola, L., Blade, C., Pujadas, G., and Salvado, M. J. (2009). *In vivo*, *in vitro*, and *in silico* studies of Cu/Zn-superoxide dismutase regulation by molecules in grape seed procyanidin extract. *Journal of Agriculture and Food Chemistry*, **57**, 3934–3942.

Ribot, J., Rodriguez, A. M., Rodriguez, E., and Palou, A. (2008). Adiponectin and resistin response in the onset of obesity in male and female rats. *Obesity*, **16**, 723–730.

Rivera, L., Morón, R., Sánchez, M., Zarzuelo, A., and Galisteo, M. (2008). Quercetin ameliorates metabolic syndrome and improves the inflammatory status in obese zucker rats. *Obesity*, **16**, 2081–2087.

Roig, R., Cascon, E., Arola, L., Blade, C., and Salvado, M. J. (1999). Moderate red wine consumption protects the rat against oxidation *in vivo*. *Life Sciences*, **64**, 1517–1524.

Roig, R., Cascón, E., Arola, L., Bladé, C., and Salvad6, M. J. (2002). Procyanidins protect fao cells against hydrogen peroxide-induced oxidative stress. *Biochimica et Biophysica Acta (BBA)–General Subjects*, **1572**, 25–30.

Schäfer, A. and Högger, P. (2007). Oligomeric procyanidins of french maritime pine bark extract (pycnogenol®) effectively inhibit α-glucosidase. *Diabetes Research and Clinical Practice*, **77**, 41–46.

Serra, A., Maci, A., Romero, M., Valls, J., Blad, C., Arola, L., and Motilva, M. (2010). Bioavailability of procyanidin dimers and trimers and matrix food effects in *in vitro* and *in vivo* models. *British Journal of Nutrition*, **103**, 944–952.

Shafrir, E., Ziv, E., and Mosthaf, L. (1999). Nutritionally induced insulin resistance and receptor defect leading to β-cell failure in animal models. *Annals of the New York Academy of Sciences*, **892**, 223–246.

Stein, J. H., Keevil, J. G., Wiebe, D. A., Aeschlimann, S., and Folts, J. D. (1999). Purple grape juice improves endothelial function and reduces the susceptibility of LDL cholesterol to oxidation in patients with coronary artery disease. *Circulation*, **100**, 1050–1055.

Storlien, L H., Higgins, J. A., Thomas, T. C., Brown, M. A., Wang, H. Q., Huang, X. F., and Else, P. L. (2000). Diet composition and insulin action in animal models. *British Journal of Nutrition*, **83**, S85–90.

Tartaglia, L. A. (1997). The leptin receptor. *Journal of Biological Chemistry*, **272**, 6093–6096.

Terra, X., Montagut, G., Bustos, M., Llopiz, N., Ardevol, A., Blade, C., Fernandez-Larrea, J., Pujadas, G., Salvado, J., Arola, L., and Blay, M. (2009). Grape-seed procyanidins prevent low-grade inflammation by modulating cytokine expression in rats fed a high-fat diet. *Journal of Nutritional Biochemistry*, **20**, 210–218.

Terra, X., Pallarés, V., Ardèvol, A., Bladé, C., Fernández-Larrea, J., Pujadas, G., Salvadó, J., Arola, L., and Blay, M. (2011). Modulatory effect of grape-seed procyanidins on local and systemic inflammation in diet-induced-obesity rats. *The Journal of Nutritional Biochemistry,* **22**(4):380–387.

Tomaru, M., Takano, H., Osakabe, N., Yasuda, A., Inoue, K., Yanagisawa, R., Ohwatari, T., and Uematsu, H. (2007). Dietary supplementation with cacao liquor proanthocyanidins prevents elevation of blood glucose levels in diabetic obese mice. *Nutrition*, **23**, 351–355.

Tran, L., Yuen, V., and McNeill, J. (2009). The fructose-fed rat: A review on the mechanisms of fructose-induced insulin resistance and hypertension. *Molecular and Cellular Biochemistry*, **332**, 145–159.

Tsai, H. Y., Wu, L. Y., and Hwang, L. S. (2008). Effect of a proanthocyanidin-rich extract from longan flower on markers of metabolic syndrome in fructose-fed rats. *Journal of Agricultural and Food Chemistry*, **56**, 11018–11024.

Varga, O., Harangi, M., Olsson, I. A. S., and Hansen, A. K. (2009). Contribution of animal models to the understanding of the metabolic syndrome: A systematic overview. *Obesity Reviews*, 99–99.

Williams, R. J., Spencer, J. P. E., and Rice-Evans, C. (2004). Flavonoids: Antioxidants or signalling molecules? *Free Radical Biology and Medicine*, **36**, 838–849.

Wolfram, S., Raederstorff, D., Preller, M., Wang, Y., Teixeira, S., Riegger, C., and Weber, P. (2006). Epigallocatechin gallate supplementation alleviates diabetes in rodents. *Journal of Nutrition*, **136**, 2512–2518.

Yokozawa, T., Kim, H. J., and Cho, E. J. (2008). Gravinol ameliorates high-fructose-induced metabolic syndrome through regulation of lipid metabolism and proinflammatory state in rats. *Journal of Agricultural and Food Chemistry*, **56**, 5026–5032.

Yuste, P., Longstaff, M., and McCorquodale, C. (1992). The effect of proanthocyanidin-rich hulls and proanthocyanidin extracts from bean (*vicia faba* L.) hulls on nutrient digestibility and digestive enzyme activities in young chicks. **67**, 57–65.

Zern, T. L. and Fernandez, M. L. (2005). Cardioprotective effects of dietary polyphenols. *Journal of Nutrition*, **135**, 2291–2294.

Zern, T. L., Wood, R. J., Greene, C., West, K. L., Liu, Y., Aggarwal, D., Shachter, N. S., and Fernandez, M. L. (2005). Grape polyphenols exert a cardioprotective effect in pre- and postmenopausal women by lowering plasma lipids and reducing oxidative stress. *Journal of Nutrition*, **135**, 1911–1917.

Zhang, H., Ji, B., Chen, G., Zhou, F., Luo, Y., Yu, H., Gao, F., Zhang, Z., and Li, H. (2009). A combination of grape seed-derived procyanidins and gypenosides alleviates insulin resistance in mice and HepG2 cells. **74**, H1–7.

6

Alani, S. R. Smith, D. M., and Markakis, P. (1989). α-galactosidases of *Vigna unguiculata. Phytochemistry*, **28** (8), 2047–2051.

Anisha, G. S. and Prema, P. (2008). Reduction of non-digestible oligosaccharides in horse gram and green gram flours using crude α-galactosidase from *Streptomyces griseoloalbus. Food Chemistry*, **106**, 1175–1179.

Annunziato, M. E., Mahoney, R. R., and Mudgett, R. E. (1986). Production of α-galactosidase from Aspergillus oryzae grown in solid state culture. *Journal of Food Science*, **51**(5), 1370–1371.

Aranda, P., Dostalova, J., Frias, J., Lopez-Jurado, M., Kozlowska, H., Pokorny, J., Urbano, G., Vidal-Valverde, C., and Zdyunczyk, Z. (2000). Nutrition. In C.L.Hedley (Ed.), *Carbohydrates in Grain Legume Seeds: Improving Nutritional Quality and Agronomic Characteristics* (pp. 61–87). Wallingford: CAB International.

Azeke, M. A., Fretzdorff, B., Buening-Pfaue, H., and Betsche, T. (2007). Comparative effect of boiling and solid substrate fermentation using the tempeh fungus (*Rhizopus oligosporus*) on the flatulence potential of African yambean (*Sphenostylis stenocarpa* L.) seeds. *Food Chemistry*, **103**(4), 1420–1425.

Becerra, M. and González Siso, M. I. (1996). Yeast β-galactosidase in solid-state fermentations. *Enzyme and Microbial Technology*, **19**, 39–44.

Biranjan, J. R., Ingole, D. R., and Sanyal, S. R. (1983). Utility of soymilk in dietary therapy of undernourished children. *Indian Medical Gazette*, **9**, 313–316.

Buchholz, K., Kasche, V., and Bornscheuer, U. T. (2005). Biocatalysts and enzyme technology. Weinheim: Wiley.

Cao, Y., Yang, P., Shi, P., Wang, Y., Luo, H., Meng, K., Zhang, Z., Wu, N., Yao, B., and Fan, Y. (2007). Purification and characterization of a novel protease-resistant α-galactosidase from *Rhizopus* sp. F78 ACCC 30795. *Enzyme and Microbial Technology*, **41**, 835–841.

Caplicec, E. and Fitzgerald, G. F. (1999). Food fermentations: role of microorganisms in food production and preservation. *International Journal of Food Microbiology*, **50**, 131–149.

Chantarangsee, M., Tanthanuch, W., Fujimura, T., Fry, C. S., and Cairns, J. K. (2007). Molecular characterization of β-galactosidases from germinating rice (*Oryza sativa*). *Plant Science*, **173**, 118–134.

Church, F. C. Jr. and Meyers, S. P. (1980). Alpha-galactosidase from pichia guilliermondii. *Mycologia*, **72**(2), 279–287.

Cruz, R., and Park, Y. K. (1982). Production of fungal a-galactosidase & its application to the hydrolysis of galactooligosaccharides in soybean milk. *Journal of Food Science*, **47**, 1973–1975.

Dagbagli, S. and Goksungur, Y. (2008). Optimization of α-galactosidase production using Kluyveromyces lactis NRRL Y-8279 by response surface methodology. *Electronic Journal of Biotechnology*, **11**(4)

De Lumen, O. B (1992). Molecular strategies to improve protein quality and reduce flatulence in legumes: a review. *Food Structure*, **11**, 33–46.

Deshpande, S. S. (1992). Food legumes in human nutrition: a personal perspective. *Critical Reviews in Food Science and Nutrition*, **32**, 333–363.

Deshpande, S. S. (2002). *Handbook of food toxicology*. New York: CRC Press, p 920.

Dey, P. M. (1985). d-Galactose-containing oligosaccharides. In P. M. Dey and R. Dixon (Eds), *Biochemistry of Storage Carbohydrates in Green Plants* (pp. 53–129). New York: Academic Press.

Dey, P. M. and Pridham, J. B. (1972). Biochemistry of α-galactosidase. *Advance Enzymology*, **36**, 911–930.

Donkor, O. N., Henriksson, A., Vasiljevic, T., and Shah, N. P. (2007). α-Galactosidase and proteolytic activities of selected probiotic and dairy cultures in fermented soymilk. *Food Chemistry*, **104**, 10–20.

Dorland's Illustrated Medical Dictionary. (2006). Available at: http://www.mercksource.com/pp/us/cns/cns_hl_dorlands.jspzQzpgzEzzSzppdocszSzuszSzcommonzSzdorlandszSzdorlandzSzdmd_g_01zPzhtm#12382991

Duszkiewicz-Reinhard, W., Gujska, E., and Khan, K. (2002). Reduction of stachyose in legume flours by lactic acid bacteria. *Journal of Food Science*, **59**(1), 115–117.

FAO (1998). Corporate Document Repository. Carbohydrates in human nutrition. *Dietary carbohydrate composition*, Available at: http://www.fao.org/docrep/W8079E/W8079E00.htm

Fialho, L. S., Guimaraes, V. M., Barros, E. G., Moreira, M. A., Dias, L. A. S., Oliveira, M. G. A., Jose, I. C., and Rezende, S. T. (2006). Biochemical composition and indigestible oligosaccharides in phaseolus vulgaris L seeds. *Plant Foods For Human Nutrition, Inglaterra*, **64**(2), 83–85.

Garro, M., Valdez, G. F., Oliver, G., and Giori, G. S. (1996). Purification of α-galactosidase from Lactobacillus fermentum. *Journal of Biotechnology*, **45**, 103–109.

Gote, M., Umalkar, H., Khan, I., and Khire, J. (2004). Thermostable α-galactosidase from Bacillus stearothermophilus (NCIM 5146) and its application in the removal of flatulence causing factors from soymilk. *Process Biochemistry*, **39**, 1723–1729.

Gote, M. M., Khan, M. I., Gokhale, D. V., Bastawde, K. B., and Khire, J. M. (2006). Purification, characterization and substrate specificity of thermostable α-galactosidase from Bacillus stearothermophilus (NCIM-5146). *Process Biochemistry*, **41**, 1311–1317.

Guimarães, V. M, de Rezende, S. T, Moreira, M. A, de Barros, E. G, and Felix, C. R. (2001). Characterization of α-galactosidases from germinating soybean seed and their use for hydrolysis of oligosaccharides. *Phytochemistry*, **58**, 67–73.

Guimarães, W. V., Dudey, G. L., and Ingram, L. O. (1992). Fermentation of sweet whey by ethanologenic Escherichia coli. *Biotechnology & Bioengineering*, **40**, 41–45.

Gul-Guven, R., Guven, K., Poli, A., and Nicolaus, B. (2007). Purification and some properties of a β-galactosidase from the thermoacidophilic *Alicyclobacillus acidocaldarius* subsp. rittmannii isolated from Antarctica. *Enzyme and Microbial Technology*, **40**, 1570–1577.

Haagenson, D. M., Klotz, K. L., and Campbell, L. (2008). Impact of storage temperature, storage duration, and harvest date on sugarbeet raffinose metabolism. *Postharvest Biology and Technology*, **49**, 221–228.

Hitz, W. D., Carlson, T. J., Kerr, P. S., and Sebastian, S. A. (2002). Biochemical and molecular characterization of a mutation that confers a decreased raffinosaccharide and phytic acid phenotype on soybean seeds. *Plant Physiology*, **128**, 650–660.

Horbowicz, M. and Obendorf, R. L. (1994). Seed desiccation tolerance and storability: dependence on flatulence-producing oligosaccharides and cyclitols—review and survey. *Seed Science Research*, **4**, 385–405.

Hsu, C. A., Yu, R. C., and Chou, C. C. (2005). Production of β-galactosidase by Bifidobacteria as influenced by various culture conditions. *International Journal of Food Microbiology*, **104**, 197–206.

Hungria, M., Andrade, D. S, Chueire, L. M. O., Probanza, A., Guttierrez-Mañero, F. J., and Megıas, M. (2000). Isolation and characterization of new efficient and competitive bean (*Phaseolus vulgaris* L.) rhizobia from Brazil. *Soil Biology and Biochemistry*. **32**, 1515–1528.

Insel, P., Ross, D., and McMahon, K. (2011). Nutrition (4th ed.). Canada: Jones and Bartlett Publishers.

Jangchud, K. and Bunnang, N. (2001). Effect of soaking time and cooking time on qualities of red kidney bean flour. *Kasetsart Journal*, **35**, 409–415.

Jones, D. A., Du Pont, M. S., Ambrose, M. J., Frias, J., and Hedley, C. L. (1999). The discovery of compositional variation for the raffinose family of oligosaccharides in pea seeds. *Seed Science Research*, **9**, 305–310.

Kestwal, R. M. and Bhide, S. V. (2007). Purification of β-galactosidase from *erythrina indica*: Involvement of tryptophan in active site. *Biochimica et Biophysica Acta*, **1770**, 1506–1512.

Khare, S. K., Jha, K., and Gandhi, A. P. (1994). Hydrolysis of flatulence-causing oligosaccharides by agarose-entrapped *Aspergillus oryzae* cells. *Food Chemistry*, **51**, 29–31.

Konsoula, Z. and Liakopoulou-Kyriakides, M. (2007). Co-production of α-amylase and β-galactosidase by *Bacillus* subtilis in complex organic substrates. *Bioresource Technology*, **98**, 150–157.

Kotwal, S. M., Gote, M. M., Sainkar, S. R., Khan, M. I., and Khire, J. M. (1997). Production of α-galactosidase by thermophilic fungus *Humicola sp.* in solid-state fermentation and its application in soyamilk hydrolysis. *Process Biochemistry*, **33**, 337–343.

Kuo, T. M., VanMiddlesworth J. F., and Wolf, W. J. (1988). Content of raffinose oligosaccharides and sucrose in various plant seeds. *Journal of Agricultural and Food Chemistry*, **36**, 32–36.

Kim, Yeong-Su, Park, Chong-Su, and Oh, eok-Kun (2006). Lactulose production from lactose and fructose by thermostable ß-lactosidase from *Sulfolobus solfataricus*. *Enzyme and Microbial Technology*, **39**, 903–908.

LeBlanc, J. G., Piard, J. C., Sesma, F., and Giori, G. S. (2005). *Lactobacillus* fermentum CRL 722 is able to deliver active alpha-galactosidase activity in the small intestine of rats. *FEMS Microbiology Letters*, **248**, 177–182.

Li, S. C., Han, J. W., Chen, K. C., and Chen, C. S. (2001). Purification and characterization of isoforms of β-galactosidases in mung bean seedlings. *Phytochemistry*, **57**, 349–359.

Lyimo M, Mugula J, and Elias T (1992). Nutritive composition of broth from selected bean varieties cooked for various periods. *Journal of the Science of Food and Agriculture*, **58**, 535–539.

Makowski, K., Bialkowska, A., Olczar, J., Kur, J., and Turkiewicz, M. (2009). Antartic, cold-adapted ß-galactosidase of *Pseudoalteronomas sp.* 22b as an effective tool for alkyl galactopyranosides synthesis. *Enzyme and Microbial Technology*, **44**, 59–64.

Manzanares, P., Graaff, L. H., Visser, J. (1998). Characterization of galactosidases from *Aspergillus niger*: Purification of a novel α-galactosidase activity. *Enzyme and Microbial Technology*, **22**, 383–390.

Mansour, E. H. and Khalil, A. H. (1998). Reduction of raffinose oligosaccharides in chickpea (*Cicer arietinum*) flour by crude extracellular fungal α-galactosidase. *Journal of the Science of Food and Agriculture*, **78**(2), 175–181.

Martínez-Villaluenga, C., Fríasa, J., Gómez R., and Vidal-Valverde, C. (2006). Influence of addition of raffinose family oligosaccharides on probiotic survival in fermented milk during refrigerated storage. *International Dairy Journal*, **16**(7), 768–774.

Mayer, J., Kranz, B., and Fischer, L. (2010). Continous production of lactulose by immobilized thermostable ß-glucosidases from *Pyrococcus furiosus*. *Journal of Biotechnology*, **145**, 387–393.

Miszkiewicz, H. and Galas, E. (2000). Removal of raffinose galactooligosaccharides from lentil (*lens culinaris* med.) by the *Mortierella vinacea* IBT-3 α-galactosidase. In: S. Bielecki, J. Tramper, and J. Polak (Eds.), *Food Biotechnology, Progress in Biotechnology* (Vol. 17). Amsterdam: Elsevier Science.

Miszkiewicz, H. and Galas, E. (2000). Removal of raffinose galactooligosaccharides from lentil (*lens culinaris* med.) by the *Mortierella vinacea* IBT-3 α-galactosidase. *Progress in Biotechnology*, **17**, 193–199.

Moriwaki, C.and Matioli, G. (2000). Influence of β-galactosidase in milk technology and lactose metabolism. *Arquivos De Ciência Da Saúde Da Unipar*, **4**(3), 283–290.

Muzquiz, M., Burbano, C., Pedrosa, M. M., Folkman, W., and Gulewicz, K. (1999). Lupins as a potential source of raffinose family oligosaccharides Preparative method for their isolation and purification. *Industrial Crops and Products*, **19**, 183–188

Novalin, S., Neuhaus, W., and Kulbe, K. D. (2005). A new innovative process to produce lactos-reduced skim milk. *Journal of Biotecnology*, **119**, 212–218.

Obendorf, R. L. (1997) Oligosaccharides and galactosyl cyclitols in seed desiccation tolerance. *Seed Science Research*, **7**, 63–74.

Omogbai, B. A., Ikenebomeh, M. J., and Ojeaburu, S. I. (2005). Microbial utilization of stachyose in soymilk yogurt production. *African Journal of Biotechnology*, **4**(9), 905–908.

Onyenekwe, P. C., Njoku, G. C., and Ameh, D. A. (2000). Effect of Cowpea (*Vigna ilwguiculata)* processing methods on flatus causing oligosaccharides. *Nutrition Research*, **20**(3), 349–358.

Ortolani, C. and Pastorello, E. A. (2006). Food allergies and food intolerances. *Best Practice & Research Clinical Gastroenterology*, **20**(3), 467–483.

Osaana, N. Donkor, A. Henriksson, T. Vasiljevic, and Shah, N. P. (2007). α-galactosidase and proteolytic activities of selected probiotic and dairy cultures in fermented soymilk. *Food Chemistry*, **104**, 10–20.

Passerat, B. and Desmaison, A. M. (1995). Lactase activity of Bifidobacterium bifidum. *Nutrition Research*, **15**(9), 1287–1295.

Pivarnik, L. F., Senacal, A. G., and Rand, A. G. (1995). Hydrolytic and transgalactosylic activities of commercial β-galactosidase (lactase) in food processing. *Advances in Food & Nutrition Research*, **38**, 1–102.

Peterbauer, T., Karner, U., Mucha, J., Mach, L., and Jones, D. A. (2003). Enzymatic control of the accumulation of verbascose in pea seeds. *Plant, Cell & Environment*, 26(8), 1385–1391.

Porter, J. E. (1992). Effect of pH on subunit association and heat protection of soybean α-galactosidase. *Enzyme Microbial Technology, New York*, **14**, 609–613.

Porzucek, H., Duszkiewicz-Reinhard, W., Piecyk, M., Klepacka, M., and Gniewosz, M. (2002). Changes of Flatulence-Causing Sugars in Legume Protein Samples by High Hydrostatic Pressure. *Food Science and Technology*, **5**(2), 07.

Potter, N. N. and Hotchkiss, (1995). *Food science* (5th ed., pp. 608). New York: Chapman & Hall.

Reddy, N. R., Salunkhe, D. K., and Sharma, R. P. (1980). Flatulence in rats following ingestion of cooked & germinated black gram & a fermented product of black gram & rice blend. *Journal of Food Science*, **45**, 1161–1164.

Rezende, S. T., Guimarães, V. M., Rodrigues, M. C., and Felix, C. R. (2005). Purification and characterization of an alpha-galactosidase from *Aspergillus fumigatus*. *Brazilian Archives of Biology and Technology*, **48**(2), 195–202.

Rezessy-Szabó, J. M., Nguyen, Q. D., Bujna, E., Takács, K., Kovács, M., and Hoschke, A. (2003). Thermomyces lanuginosus CBS 395.62/b Strain as Rich Source of α-Galactosidase Enzyme, *Food Technology and Biotechnology*, **41**(1), 55–59.

Rhimi, M., Aghajari, M., Jaouadi, B., Juy, M., Boudebbouze, S., Maguin, E., HAser, R., and Bejar, S. (2006). Exploring the acidotolerance of ß-lactose from *Lactobacillus delbrueckii* subsp *bulgaricus*: an attractive enzyme for lactose convertion. *Research in Microbiology*, **160**, 775–784.

Rojas, A. L., Nagem, R. A. P., Neustroev, K. N., Arand, M., Adamska, M., Eneyskaya, E. V., Kulminskaya, A. A., Garratt, R. C., Golubev, A. M., Polikarpov, I., and Rojas, A. L. (2004). Crystal structures of β-Galactosidases from *Penicillium* sp. and its complex with galactose. *Journal of Molecular Biology*, **343**(5), 1281–1292.

Salunkhe, D. K. and Kadam, S. S. (1989). *Hand book of world legumes: Nutritional chemistry processing technology and utilization*, Boca Raton, Florida: CRC Press Inc.

Sanada, C. T. N., Karp, S. G., Spier, M. R., Portella, A. C., Gouvêa, P. M, Yamaguishi, C. T., Vandenberghe, L. P. S., Pandey, A., and Soccol, C. R. (2009). Utilization of soybean vinasse for α-galactosidase production. *Food Research International*, **42**, 476–483.

Santiago, P. A., Marquez, L. D. S., Cardoso, L. D. S., and Ribeiro, E. J. (2004). Estudo da Produção de β-galactosidase por fermentação de soro de queijo com Kluyveromyces marxianus. *Ciência e Tecnologia de Alimentos*, **24**(4), 567–572.

Scalabrini, P., Rossi, M., Spettoli, P., and Matteuzzi, D. (1998). Characterization of bifidobacterium strains for use in soymilk fermentation. *International Journal of Food Microbiology*, **39**, 213–219.

Shankar, S. K. and Mulimani, V. H. (2007). α-Galactosidase production by *Aspergillus oryzae* in solid-state fermentation. *Bioresource Technology*, **98**, 958–961.

Shivanna, B. D., and Ramakrishan, M. (1985). α-Galactosidase from germinating guar (*Cyamopsis tetragonolobus*). *Journal of Biosciences*, **9**, 109–116.

Shurtleff, W. and Aoyagi, A. (1999). *The book of tofu*, Tokyo.

Soh, C. P., Ali, Z. M., and Lazan, H. (2006). Characterisation of an α-galactosidase with potential relevance to ripening related texture changes. *Phytochemistry*, **67**, 242–254.

Song, D., Chang, S. K. C., and Ibrahim, S. A. (2009). Descriptive sensory characteristics of no flatulence pinto bean. *Journal of Food Quality*, doi 10.1111/j.1745-4557.2009.00278.

Spier, M. R., Habu, S., Scheidt, G. N., Woiciechowski, A. L., Pandey, A., and Soccol, C. R. (in press). Fermented functional foods. In: *Comprehensive Food Fermentation Biotechnology*, Chapter 29.

Thippeswamy, S. and Mulimani, V. H. (2002). Enzymic degradation of raffinose family oligosaccharides in soymilk by immobilized a-galactosidase from Gibberella fujikuroi. *Process Biochemistry*, **38**, 635–640.

Tsangalis, D. and Shah, N. P. (2004). Development of an isoflavone aglycone-enriched soymilk using soy germ, soy protein isolate and bifidobacteria, *Food Research International*, **37**(4), 301–312.

Tzortzis, G., Jay, A. J., Baillon, M. L. A., Gibson, G. R., and Rastall, R. A. (2003). Synthesis of α-galactooligosaccharides with α-galactosidase from Lactobacillus reuteri of canine origin. *Applied Microbiology and Biotechnology*, **63**(3), 1432–1614.

Vasiljevic, T., and Jelen, P. (2001). Production of β-galactosidase for lactose hydrolysis in milk and dairy products using thermophilic lactic acid bacteria. *Innovative Food Science and Emerging Technologies*, **2**(2), 75–85.

Viana, P. A., de Rezende, S. T., Alves, A. A., Manfrini, R. M., Alves, R. J., Bemquerer, M. P., Santoro, M. M., and Guimarães, V. M. (2011). Activity of Debaryomyces hansenii UFV-1 α-galactosidases against α-Dgalactopyranoside derivatives. *Carbohydrate Research*, doi: 10.1016/j.carres.2011.01.024.

Viana S. F., Guimarães V. M., Jose I. C., Oliveira M. G. A., Costa, N. M. B., de Barros, E. G., Moreira, M. A., and de Rezende, S. T. (2005). Hydrolysis of oligosaccharides in soybean flour by soybean α-galactosidase. *Food Chemistry*, **93**(4), 665–670.

Wiseman, A. (1991). *Manual de biotecnología de los enzimas*. Zaragoza: Acribia.

Yamaguishi, C. T., Sanada, C. T., Gouvêa, P. M., Pandey, A., Woiciechowski, A. L., Parada, J. L., and Soccol, C. R. (2009). Biotechnological process for producing black bean slurry without stachyose. *Food Research International*, **42** (4), 425–429.

Zadow, J. G. (1993). Economic considerations related to the production of lactose and lactose by-products. *IDF Bulletin*, **289**, 10.

7

Agrahar-murugkar, D. and Subbulakshmi, G. (2005). Nutritional value of edible wild mushrooms collected from the Khasi hills of Meghalaya. *Food Chemistry*, **89**, 599–603.

Aida, f. M. N. A., Shuhaimi, M., Yazid, M., and Maaruf, A. G. (2009). Mushroom as a potential source of prebiotics: A review. *Trends in Food Science and Technology*, doi: 10,1016/j.tifs.2009.07.007.

Bao, X., Duan, J., Fang, X., and Fang, J. (2001). Chemical modifications of the (1-3)-α-D-glucan from sporos of *Ganoderma lucidum* and investigation of their physicochemical properties and immunological activity. *Carbohydrate Research*, **336**, 127–140.

Barros, L., Cruz, T., Baptista, P., Estevinho, L. M., and Ferreira, I. C. F. R. (2008a). Wild and commercial mushrooms as source of nutrients and nutraceuticals. *Food and Chemical Toxicology*, **46**, 2742–2747.

Barros, L., Falcão, S., Baptista, P., Freire, C. Vilas-boas, M., and Ferreira, I. C. F. R. (2008b). Antioxidant activity of *Agaricus* sp Mushroom by chemical, biochemical and eletrochemical assays. *Food Chemistry*, **111**, 61–66.

Bhat, T. A. and Singh, R. P. (2008). Tumor angiogenesis—A potential target in cancer chemoprevention. *Food and Chemical Toxicology*, **46**, 1334–1345.

Bobek, P., Lubomir, O., and Galbavy, S. (1998). Dose and time dependent hypocholesterolemic effect of oyster mushroom (*Pleurotus ostreatus*) in rats. *Basic Nutritional Investigation*, **14** (3), 282–286.

Carbonero, E. R., Gracher, A. H. P., Smiderle, F. R., Rosado, F. R., Sassaki, G. L., Gorin, P. A. J., and Iacomini, M. A. (2006). b-glucan from the fruit bodies of edible mushrooms *Pleurotus eryngii* and *Pleurotus ostreatoroseus*. *Carbohydrate Polymers*, **66**, 252–257.

Chattopadhyay, N., Ghosh, T., Sinka, S., Chattopadhyay, K., Karmakar, P., and Ray, B. (2009). Polysaccharides from *Turbinaria conoides*: Structural features and antioxidant capacity. *Food Chemistry*, doi: 10.1016/jfood-chem.200905.069.

Cheng, J.-J., Lin, C-Y., Lur, H. S., Chen, H. P., and Lu, M. K. (2008). Properties and biological functions of polysaccharides and ethanolic extracts isolated from medicinal fungus, *Fomitopsis pinicola*. *Process Biochemistry*, **43**, 829–834.

Chen, J. and Seviour, R. (2007) Medicinal importance of fungal ß-(1→3), (1→6)-glucans. *Mycological Research*, **111**, 635–652.

Chen, W., Zhao, Z., Chen, S-F., and Li, Y-Q, (2008). Optimization for the productionof exopolysaccharide from *Fomes fomentarius* in submerged culture and its antitumor effect *in vitro*. *Biosource Technology*, **99**, 3187–3194.

Cheung, P. C. K. (1996). The hypocholesterolemic effect of two edible mushrooms: *Auricularia auricula* (Tree-ear) and *Tremella fuciformes* (White Jelly-leaf) in hypercholesterolemic rats. *National Research*, **6**(10), 1721–1725.

Choi, Y., Lee, S. M., Chun, J., Lee, H. B., and Lee, J. (2006). Influence of heat treatment on the antioxidant activities and polyphenolic compounds of Shiitake (*Lentinus edodes*) mushroom. *Food Chemistry*, **99**, 381–387.

Contois, L., Akalu, A., and Brooks, P. C. (2009). Integrins as "functional hubs" in the regulation of pathological angiogenesis. *Seminars in Cancer Biology*, doi:10.1016/j.semcancer.2009.05.002

Cui, F. J., Tao, W. Y., Xu, Z. H., Guo, W. J., Xu, H. Y., Ao, Z. H., Jin, J., and Wei, Y. Q. (2007). Structural analysis of anti-tumor heteropolysaccharide GFPS1b from the cultured mycelia of *Grifola frondosa* GF9801. *Bioresource Technology*, 98, 395–401.

Diyabalanage, T., Mulabagal, V., Mills, G., DeWitt, D. L., M., and Nair, M. G. (2008). Health-beneficial qualities of the edible mushroom. *Agrocybe aegerita*. *Food Chemistry*, **108**, 97–102.

Dore, C. M. P.G., Azevedo, T. C. G., Souza, M. C. R., Rego, L. A., Dantas, J. C. M., Silva, F. R.F., Rocha, H. A. O., Baseia, I. G., and Leite, E. L. (2007). Anti-inflammatory, antioxidant and cytotoxic actions of ß-glucan-rich extract from *Geastrum saccatum* mushroom. *International Immunopharmacology*, **7**, 1160–1169.

Duarte, M., Longatto Filho, A., and Schmitt, F. C. (2007). Angiogenesis, haemostasis and cancer: New paradigms old concerns. *Jornal Brasileiro de Patologia e Medicina Laboratorial*, **43**(6), 441–449.

Dudhgaonkar, S., Thyagarajan, A., and Sliva, D. (2009). Suppression of the inflammatory response by triterpenes isolated from the mushroom *Ganoderma lucidum*. *International Immunopharmacology*, 2009, doi: 10.1016/j.intimp. 07.011

Elmastas, M., Isildak, O., Turkekul, I., and Temur, N. (2007). Determination of antioxidant activity and antioxidant compounds in wild edible mushrooms. *Journal of Food Composition and Analysis*, **20**, 337–345.

Fan, L., Soccol, A. T., Pandey, A., and Soccol, C. R. (2007). Effect of nutrional and environmental conditions on the production of exo-polysaccharide of *Agaricus brasiliensis* by submerged fermentation and its antitumor activity. *LTW*, **40**, 30–35.

Frank, J. A., Xiao, R., Yu, S., Fergunson, M., Hennings, L. J., Simpson, P. M., Ronis, M.J. J., Fang, N. Badger, T. M., and Simmen, F. A. (2006). Effect of Shiitake mushroom dose on colon tumorigenisis in azoxymethane-treatd male Sprague-Dawley rats. *Nutrition Research*, **26**, 138–145.

Ghorai, S., Banik, S. P., Verma, D., Chowdhury, S., Mukherjee, S., and Khowala, S. (2009). Fungal biotechnology in food and feed processing. *Food Research International*, **42**, 577–587.

Gu, Y-H. and Belury, M. A. (2005). Selective induction of apoptosis in murine skin carcinoma cells (CH72) by an ethanol extract of *Lentinula edodes*. *Cancer Letters*, 220, 21–28.

Guo, Y., Wang, H., and Ng, T. B. (2005). Isolation of trichogin, an antifungal protein from fresh fruiting bodies of the edible mushroom *Tricholoma giganteum*. *Peptides*, **26**, 575–580.

Hagiwara, S., Takahashi, M., Shen, Y., Kaihou, S., Tomiyama, T., Yazawa, M., Tamai,Y., Sin, Y., Kazusaka, A., and Terezawa, M. A. (2005). Phytochemical in the edible Tamogi-take mushroom (*Pleurotus cornucopiae*), D-manitol, inhibits

ACE activity and lowers the blood pressure of spontaneously hypertensive rats. *Bioscience, Biotechnology, and Biochemistry*, **69**(8), 1603–1605.

Han, C., Yuan, J., Wang, Y., and Li, L. (2006). Hypoglycemic activity of fermented mushroom of *Coprinus comatus* rich in vanadium. *Journal of Trace Elements in Medicine and Biology*, **20**, 191–196.

Harhaji, Lj., Mijatovic, S., Maksimovic-Ivanic, D., Stojanovic, I., Momcilovic, M., Maksimovic, V., Tufegdzie, S., Marjanovic, Z., Mostarica-Stojkovic, M., Vucinic, Z., and Stosic-Grujicic, S. (2008). Antitumor effect of *Coriolus versicolor* methanol extract against mouse B16 melanoma cells: *in vitro* and *in vivo* study. *Food and Chemical Toxicology*, **46**, 1825–1833.

Hatvani, N. (2001). Antibacterial effect of the culture fluid of *Lentinus edodes* mycelium grown in submerged liquid culture. *International Journal of antimicrobial Agents*, **17**, 71–74.

Hearst, R., Nelson, D., McCollum, G., Millar, B. C., Maeda, Y., Goldsmith, C. E., Rooney, P. J., Loughrey, A., Rao, J. R., and Moore, J. E. (2009). An examination of antibacterial and antifungical properties of constituents of Shiitake (*Lentinula edodes*) and Oyster (*Pleurotus ostreatus*) mushrooms. *Complementary Therapies in Clinical Practice*, **15**, 5–7.

Hirasawa, M., Shouji, N., Neta, T., Fukushima, K., and Takada, K. (1999). Three kinds of antibacterial substances from *Lentinus edodes* (Berk.) Sing. (Shiitake, an edible mushroom). *International Journal of antimicrobial Agents*, **11**, 151–157.

Hsu, S-C., Ou, C-C., Chuang, T-C., Li, J-W., Lee, Y-J., Wang, V., Liu, J-Y., Chen, C-S., Lin, S-C., and Kao, M-C. (2009). *Ganoderma tsugae* extract inhibits expression of epidermal growth factor receptor and angiogenesis in human epidermoid carcinoma cells: *in vitro* and *in vivo*. *Cancer Letters*, **28**, 108–116.

Huang, Q., Jin, Y., Zhang, L., Cheung, P. C. K., and Kennedy, J. F. (2007). Structure, molecular size and antitumor activities of polysaccharides from *Poria cocos* mycelia produced in fermented. *Carbohydrate Polymers*, **70**, 324–333.

Hwang, H-J., Kim, S-W., Lim, J-M., Joo, J-H., Kim, H-O., Kim, H-M., and Yun, J-W. (2005). Hypoglicemic effect of crude exopolysaccharides produced by a medicinal mushroom *Phellinus baumii* in streptozotocin-induced diabetic rats. *Life Science*, **76**, 3069–3080.

Jayakumar, T., Thomas, P. A., and Geraldine, P. (2009). *In vitro* antioxidant activities of an ethanolic extract of the oyster mushroom *Pleurotus ostreatus*. *Innovative Food Science and Emerging Technologies*, **10**, 228–234.

Ji, Z., Tang,Q., Zhang, J., Yang, Y., Jia, W., and Pan, Y. (2007). Immunomodulation of RAW264.7 macrophages by GLIS, a proteopolysaccharide from *Ganoderma lucidum*. *Journal of Ethnopharmacology*, **112**, 445–450.

Jung, H-J., Kang, H-J., Song, Y. S., Park, E-H., Kim, Y-M., and Lim, C-J. (2008). Anti-inflammatory, anti-angiogenic and anti-nociceptive activities of *Sedum sarmentosum* extract. *Journal of Ethnopharmacology*, **116**, 138–143.

Kalac, P. (2009). Chemical composition and nutritional value of European species of wild growing mushrooms: A review. *Food Chemistry*, **113**, 9–16.

Kasai, H., He, L. M., Kawamura, M., Yang, P.T., Deng, X. W., Munkanta, M., Yamashita, A., Terunuma, H., Hirama, M., Horiuchi, I., Natori, T., Koga, T., Amano, Y., Yamaguchi, N., and Ito, (2004). M. Il-12 Production Induced by *Agaricus blazei* Fraction H (ABH) Involves Toll-like Receptor (TLR). *eCAM*, **1**(3), 259–267.

Kim, C-F., Jiang, J-J., Leung, K-L., Fung, K-P., and Lau, C. B-S. (2009). Inhibitory effects of *Agaricus blazei* extracts on human myeloid leukemia cells. *Journal of Ethnopharmacology*, **122**, 320–326.

Kim, Y. O., Park, H. W., Kim, J. H., Lee, J. Y., Moon, S. H., and Shin, C. S. (2006). Anti-cancer effect and structural characterization of endo-polysaccharide from cultivated mycelia of *Inonotus obliquus*. *Life Sciences*, **79**, 72–80.

Kim, S-H., Song, Y-S., Kim, S-K., Kim, B-C., Lim, C-J., and Park, E-H. (2004). Anti-inflammatory and related pharmacological activities of the n-BuOH subfraction of mushroom *Phellinus linteus*. *Journal of Ethnopharmacology*, **93**, 141–146.

Kimura, N., Fujino, E., Urabe, S., Mizutani, H., Sako, T., Imai, S., Toyoda, Y., and Arai, T.

(2006). Effect of supplementation of *Agaricus* mushroom meal extracts on enzyme activities in peripheral leukocytes of calves. *Research in Veterinary Science*, doi:10.1016/j.rvsc.2006.02.003.

Kokean, Y., Nishi, T., Sakakura, H., and Furuichi, Y. (2005). Effect of frying with edible oil on antihypertensive properties of Hatakeshimeji (*Lyophyllum decastes* Sing.) Mushroom. *Food Science Technology Research*, **11**(3), 339–343.

Kuo, M-C., Weng, C-Y., Hab, C-L., and Wu, M-J. (2006). *Ganoderma lucidum* mycelia enhance innate immunity by activating NF-B. *Journal of Ethnopharmacology*, **103**, 217–222.

Kwon, A-H., Qiu, Z., Hashimoto, M., Yamamoto, K., and Kimura, T. (2009). Effects of medicinal mushroom (*Sparassis crispa*) on wound healing in streptozotocin-induced diabetics rats. *The American Journal of Surgery*, **197**(4), 503–509.

Lee, B. C., Bae, J. T., Pyo, H. B., Choe, T. B., Kim, S. W., Hwang, H. J., and Yun, J. W. (2003). Biological activities of the polysaccharides produced from submerged culture of the edible Basidiomycete *Grifola frondosa. Enzyme and Microbial Technology*, **32**, 574–581.

Lee, Y-L., Jian, S-Y., Lian, P-Y., and Mau, J-L. (2008). Antioxidant properties of extracts from a white mutant of the mushroom *Hypsizigus marmoreus*. *Journal of Food Composition and Analysis*, **21**, 116–124a.

Lee, I. P., Kang, B.H., Roh, J. K., and Kim, J. R. (2008). Lack of carcinogenicity of lyophilized *Agaricus blazei* Murill in a F344 rat two year bioassay. *Food and Chemical Toxicology*, **46,** 87–95c.

Lee, I-K., Kim, Y-S., Jang, Y-W., Jung, J-Y., and Yun, B-S. (2007). New antioxidant polyphenols from medicinal mushroom *Inonotus obliquus*. *Bioorganic and Medicinal Chemistry Letters*, **17**, 6678–6681b.

Lee, J-S., Park, B. C., Ko, Y. J., Choi, M. K., Choi, H. G., Yong, C. S., Lee, J-S., and Kim, J-A. (2008). *Grifola frondosa* (Maitake Mushroom) water extract inhibits vascular endothelial growth factor-induced angiogenesis through inhibition of reactive oxygen species and extracellular signal-regulated kinase phosphorylation. *Journal of Medicinal Food*, **11**(4), 643–651b.

Lee, C-L., Yang, X., and Wan, J. M-F. (2006). The culture duration affects the immunomodulatory and anticancer effect of polysaccharopeptide derived from *Coriolus versicolor*. *Enzyme and Microbial Technology*, **38**, 14–21.

Lee, Y-L., Yen, M-T., and Mau, J-L. (2007). Antioxidant properties of various extracts from *Hypsizigus marmoreus*. *Food Chemistry*, **104**, 1–9a.

Leung, P. H., Zhao, S., Ho, K. P., and Wu, J. Y. (2009). Chemical properties and antioxidant activity of polysaccharides from mycelial culture of *Cordyceps sinensis* fungus Cs-HK1. *Food Chemistry*, **114**, 1251–1256.

Li, Y. R., Liu, Q. H., Wang, H. X., and Ng, T. B. (2008b). A novel lectin with potent antitumor, mitogenic and HIV-1 reverse transcriptase inhibitory activities from edible mushroom *Pleurotus citrinopileatus*. *Biochimica et Biophysica Acta*, **1780**, 51–57b.

Li, S-G., Wang, D-G., Tian, W., Wang, X-X., Zhao, J-X., Liu, Z., and Chen, R. (2008a). Characterization and anti-tumor activity of a polysaccharide from *Hedysarum polybotrys* Hand.-Mazz. *Carbohydrate Polymers*, 344–350a.

Li, S. P., Zhang, G. H., Zeng, Q., Huang, Z. G., Wang. Y. T., Dong, T. T. X., and Tsim, K. W. K. (2006). Hypoglycemic activity of polysaccharide, with antioxidation, isolated from cultured *Cordyceps* mycelia. *Phytomedicine*, **13**, 428–433a.

Lima, L. F. O., Habu, S., Gern, J. C., Nascimento, B. M., Parada, J. L., Noseda, M. D., Gonçalvez, A. G., Nisha, V. R., Pandey, A., Soccol, V. T., and Soccol, C. R. (2008). Production and Characterization of the Exopolysaccharides Produced by *Agaricus brasiliensis* in Submerged Fermentation. *Appl Biochem Biotechnol*, doi 10.1007/s12010-008-8187-2.

Liu, F., Ooi, E. C., and Chang, S. T. (1997). Free radical scavening activities of mushroom polysaccharide extracts. *Life Sciences*, **60**, 763–771.

Liu, D., Zheng, H., Liu, Z., Yang, B., Tu, W., and Li, L. (2009). Chemicalcomposition and antimicrobial activity of essential oil isolated from the culture mycelia of *Ganoderma japonicum*. *Journal of Nanjing Medical University*, **23**(3), 168–172.

Lo, H-C., Tsai, F. A., Wasser, S. P., Yang, J-G., and Huang, B. M. (2006). Effects of ingested fruiting bodies, submerged culture biomass, and acidic polysaccharide glucuronoxylomannan of *Tremella mesenterica* Retz.: Fr. On glycemic responses in normal and diabetics rats. *Life Sciences*, **78**, 1957–1966.

Lu, M-K., Cheng, J-J., Lin, C-Y., and Chang, C-C. (2009). Purification structural elucidation, and anti-inflammatory effect of a water-soluble 1, 6-branched 1, 3-α-d-galactan from cultured mycelia of *Poria cocos*. *Food Chemistry*, doi: 10.1016/j.foodchem.2009.04.126.

Lv, H., Kong, Y., Yao, Q., Zhang, B., Leng, F-W., Bian, H-J., Balzarini, J., Damme, E.V., and Bao, J-K. (2009). Nebrodeolysin, a novel hemolytic protein from mushroom *Pleurotus nebrodensis* with apoptosis-inducing and anti-HIV-1 effects. *Phytomedicine*, **16**, 198–205.

Maiti, S., Bhutia, S. K., Mallick, S. K., Niyati Khadgi, A. K., and Maiti, T. K. (2008). Antiproliferative and immunostimulatory protein fraction from edible mushrooms. *Environmental Toxicology and Pharmacology*, **26**, 187–191.

Makrilia, N., Lappa, T., Xyla, V. Nikolaidis, I., and Syrigos, K. (2009). The role of angiogenesis in solid tumours. *Europe Journal International Medicine*, doi:10.1016/j.ejim.2009.07.009.

Michelot, D. and Melendez-Howell, L. M. (2003). Amanita muscaria: Chemistry, biology, toxicology, and ethnomycology. *Mycol. Res.* **107**(2), 131–146.

Miyazawa, N., Okazaki, M., and Ohga, S. (2008). Antihypertensive effect of *Pleurotus nebrodensis* in spontaneously hypertensive rats. *Journal of Oleo Science*, **57**(12), 675–681.

Moradali, M-F., Mostafavi, H., Ghods, S., and Hedjaroude, G-A. (2007). Immunomodulating and anticancer agents in the realm of macromycetes fungi (macrofungi). *International Immunopharmacology*, **7**, 701–724.

Ngai, P. H. K. and Ng, T. B. (2004a). A mushroom (*Ganoderma capense*) lectin with spectacular thermostability, potent mitogenic activity on splenocytes and antiproliferative activity toward tumor cells. *Biochemical and Biophysical Research Communications*, **314**, 988–993.

Ngai, P. H. K. and Ng, T. B. (2004b). A ribonuclease with antimicrobial, antimitogenic and antiptoliferative activities from edible mushroom *Pleurotus sajor-caju*. *Peptides*, **25**, 11–17.

Ngai, P. H. K., Zhao, Z., and Ng, T. B. (2005). Agrocybin, an antifungal peptide from the edible mushroom *Agrocybe cylindracea*. *Peptides*, **26**, 191–196.

Park, S. E., Yoo, H. S., Jin, C-Y., Hong, S-H., Lee, Y-W., Kim, B. W., Lee, S-H., Kim, W-J., Cho, C-K., and Choi, Y-H. (2009). Induction of apoptosis and inhibition of telomerase activity in human lung carcinoma cells by the water extract of *Cordyceps militaris*. *Food and Chemical Toxicology*, **47**, 1667–1675.

Peng, Y., Zhang, L., Zeng, F., and Kennedy, J. F. (2005). Structure and antitumor activities of the water-soluble polysaccharides from *Ganoderma tsugae* mycelium. *Carbohydrate polymers*, **59**, 385–392.

Peng, Y., Zhang, L., Zeng, F., and Xu, Y. (2003). Structure and antitumor activity of extracellular polysaccharides from mycelium. *Carbohydrate Polymers*, **54**, 297–303.

Ramjaum, A. R. and Hodivala-Dilke, K. (2009). The role of cell adhesion pathways in angiogenesis. *The International Journal of Biochemistry and Cell Biology*, **41**, 521–530.

Rao, J. R., Millar, B. C., and Moore, J. E. (2009). Antimicrobial properties of shiitake mushroom (*Lentinula edodes*). *International Journal of Antimicrobial Agents*, **33**, 591–592.

Reid, G. (2008). Probiotics and prebiotics, progress and challenges. *International Dairy Journal*, doi: 10.1016/j.idairyj.2007.11.025.

Ren, G. Zhao, Y-P, Yang, L., and Fu, C-X. (2008). Anti-proliferative effect of clitocine from the mushroom *Leucopaxillus giganteus* on human cervical cancer HeLa cells by inducing apoptosis. *Cancer Letters*, **262**, 190–200.

Ribatti, D. (2009). Endogenous inhibitors of angiogenesis A historical. *Leukemia Research*, **33**, 638–644.

Ribeiro, B., Lopes, R., Andrade, P. B., Seabra, R. M., Gonçalves, R. F., Baptista, P., Quelhas, I., and Valentão, P. (2008). Comparative study of phytochemicals and antioxidant potential of wild edible mushroom caps and stipes. *Food Chemistry*, **110**, 47–56.

Roglic, G., Unwin, N., Bennett, P. H., Mathers, C., Tuomilehto, J., Nag, S., Connolly, V., and King, H., (2005, September). The burden of mortality attributable to diabetes. *Diabetes Care*, **28**(9), 2130–2135.

Rubel, R., Dalla Santa, H. S., Fernandes, L. C., Lima Filho, J. H. C., Figueiredo, B. C., DI Bernardi, R., Moreno, A. N., and Leifa, F., Soccol, C. R. (2008). High immunomodulatory and preventive effects against sarcoma 180 in mice feed with lingzhi or reishi mushroom *Ganoderma lucidum* (W. Curt.: Fr.) P. Karst. (Aphyllophoromycetideae) Mycelium. *International Journal of Medicinal Mushrooms*, **10**, 37–48.

Sadava, D. Still, D. W., Mudry, R. R., and Kane, S. E. (2009). Effect of *Ganoderma* on drug-sensitive and multidrug-resistant small-cell lung carcinoma cells. *Cancer Letters*, **277**, 182–189.

Sarangi, I., Ghosh, D., Bhutia, S. K., Mallick S. K., and Maiti, T. K. (2006). Anti-tumor and immunomodulating effects of *Pleurotus ostreatus* mycelia-derived proteoglycans. *International Immunopharmacology*, **6**, 1287–1297.

Sarikurkcu, C., Tepe, B., and Yamac, M. (2008). Evaluation of the antioxidant activity of four edible mushrooms from the Central Anatolia, Eikisehir - Turkey: *Lactarius deterrimus, Suillus collitinus, Boletus edulis, Xerocomus chrysenteron. Biosource Technology*, **99**, 6651–6655.

Seto, S. W., Lam, T. Y., Tam, H. L., Au, A. L. S., Chan, S. W., WU, J. H., YU, P. H. F., Leung, G. P. H., Ngai, S. M., Yeung, J. H. K., Leung, P. S., Lee, S. M. Y., and Kwan, Y. W. (2009). Novel hypoglycemic effects of *Ganoderma lucidum* water-extract in obese/diabetic (*+db/+db*) mice. *Phytomedicine*, **16**, 426–436.

Shih, I-L., Chou, B-W., Chen, C-C., Wu, J-W., and Hsieh, C. (2008). Study of mycelial growth and bioactive polysaccharide production in batch and fed-batch culture of *Grifola frondosa. Bioresource Technology*, **99**, 785–793.

Shimizu, S., Kitada, H., Yokota, H., Yamakawa, J., Murayama, T., Sugiyama, K., Izumi, H., and Yamaguchi, N. (2002). Activation of the alternative complement pathway by *Agaricus blazei* Murill. *Phytomedicine*, **9**, 536–545.

Shu, C-H. and Lung, M-Y. (2008). Effect of culture pH on the antioxidant properties of *Antrodia camphorata* id submerged culture. *Journal of the Chinese Institute of Chemical Engineers*, **39**, 1–8.

Sicoli, G., Rana, G. L., Marino, R., Sisto, D., Lerario, P., and Luisi, N., (2005). Forest fungi as bioindicators of a healthful environment and as producers of bioactive metabolites useful for therapeutic purposes. 1st European Cost E39 Working Group 2 Workshop: *"Forest Products, Forest Environment and Human Health: Tradition, Reality, and Perspectives."* April 20th–22nd, Christos Gallis (Ed.), Firenze, Italy.

Smirdele, F. R., Olsen, L. M., Carbonero, E. R., Baggio, C. H., Freitas, C. S., Marcon, R., Santos, A. R. S., Gorin, P. A. J., and Iacomini, M. (2008). Anti-inflammatory and analgesic properties in a rodent modelo f a (1→3), (1→6)-liked ß-glucan isolated from *Pleurotus pulmonarius. European Journal of Pharmacology*, **597**, 86–91.

Soares, A. A., Souza, C. G. M., Daniel, F. M., Ferrari, G. P., Costa, S. M. G., and Peralta, R. M. (2009). Antioxidant activity and total phenolic content of *Agaricus brasiliensis* (*Agaricus blazei* Murril) in two stages of maturity. *Food Chemistry*, **112**, 775–781.

Song, Y. S., Kim, S-H., Sa, J-H., Jin, C., Lim, C-J., and Park, E-H. (2003). Anti-angiogenic, antioxidant and xanthine oxidase inhibition activities of the mushroom *Phellinus linteus*. *Journal of Ethnopharmacology*, **88**, 113–116.

Song, T-Y., Lin, H-C., Yang, N-C., and Hu, M-L. (2008). Antiproliferative and antimetastatic effects of the ethanolic extract of *Phellinus igniarius* (Linnearus: Fries) Quelet. *Journal of Ethnopharmacology*, **115**, 50–56.

Stupack, D. G., Storgard C. M., and Cheresh, D. A. (1999). A role for angiogenesis in rheumatoid arthritis. *Brazilian Journal of Medical and Biological Research,* **32,** 573–581.

Surenjav, U., Zhang, L., Xu, X., Zhang, X., and Zeng, F. (2006). Effects of molecular structure on antitumor activities of (1-3)-b-D-glucans from different *Lentinus edodes. Carbohydrate Polymers*, **63**, 97–104.

Synytsya, A., Mickvá, K., Synytsya, A., Jablonsky, I., Spevacek, J., Erban, V., Kovarikova, E., and Copikova, J. (2009). Glucans from fruit bodies of cultivated mushrooms *Pleurotus ostreatus* and *Pleurotus eryngii*: Structure and potential prebiotic activity. *Carbohydrate Polymers*, **76**, 548–556.

Tong, H., Xia, F., Feng, K., Sun, G., Gao, X., Sun, L., Jiang, R., Tian, D. and Sun, X. (2009). Structural characterization and in vitro antitumor activity of a novel polysaccharide isolated from the fruiting bodies of *Pleurotus ostreatus*. Bioresource *Technology*, **100**, 1682–1686.

Tsai, S-Y., Huang, S-J., LO, S-H., WU, T-P., Lian, P-L., and Mau, J-L. (2008). Flavour components and antioxidant properties of several cultivated mushrooms. *Food Chemistry*, doi: 10.1016/j.foodchem.2008.08.034.

Tsai, S-Y., Tsai, H-L., and Mau, J-L. (2007). Antioxidant properties of *Agaricus blazei*, *Agrocybe cylindracea*, and *Boletus edulis*. *LWT*, **40**, 1392–1402.

Turkoglu, A., Duru, M. E., Mercan, N., Kivrak, I., and Gezer, K. (2007). Antioxidant and antimicrobial of *Laetiporus sulphureus* (Bull.) Murril. *Food Chemistry*, **101**, 267–273.

Van, Q., Nayak, B. N., Reimer, M., Jones, P. J. H., Fulcher, R. G., and Rempel, C. B. (2009). Anti-inflammatory effect of *Inonotus obliquus*, *Polygala senega* L., and *Viburnum tribolum* in a cell screening assay. *Journal of Ethnopharmacology*, doi: 101016/j.jep.2009.06.026.

Vasiljevic, T. and Shah, N. P. (2008). Probiotics - From Metchnikoff to bioactives. *International Dairy Journal*, **18**, 714–728.

Wang, Y. (2009). Prebiotics: Present and future in food science and technology. *Food Research International*, **42**, 8–12.

Wang, H. X. and Ng, T. B. (2004). Purification of a novel low-molecular-mass laccase with HIV-1 reverse transcriptase inhibitory activity from the mushroom *Tricholoma giganteum*. *Biochemical and Biophysical Research Communications*, **315**, 450–454.

Wang, H. and Ng, T. B. (2006). Ganodermin, an antifungical protein from fruiting bodies of medicinal mushroom *Ganoderma lucidum*. *Peptides*, **27**, 27–30.

Wang, J., Wang, T. B., and Ng, T. B. (2007). A peptide with HIV-1 reverse transcriptase inhibitory activity from the medicinal mushroom *Russula paludosa*. *Peptides*, **28** 560–565.

Wu, M-J., Cheng, T-L., Cheng, S-Y., Lian, T-W., Wang, L., and Chiou, S-Y. (2006). Immunomodulatory Properties of *Grifola frondosa* in Submerged Culture. *Journal of Agricultural and Food Chemistry*, **54**, 2906–2914.

Wu, P-Q., Xie, Y-Z., Li, S-Z., La Pierre, D. P., Deng, Z., Chen, Q., Li, C., Zhang, Z., Guo, J., Wong, C-K. A., Lee, D. Y., Yee, A., and Yang, B. B. (2006). Tumour cell adhesion and integrin expression affected by *Ganoderma lucidum*. *Enzyme and Microbial Technology*, **40**, 32–41.

Wu, J. Y., Zhang, Q. X., and Leung, P. H. (2006). Inhibitory effects of ethyl acetate extract of *Cordyceps sinensis* mycelium on various cancer cells in culture and B16 melanoma in C57BL/6 mice. *Phytomedicine*, 14(1), 43–49.

Xie, G., Schepetkin, I. A., and Quinn, M. T. (2007). Immunomodulatory activity of acidic polysaccharides isolated from *Tanacetum vulgare* L. *International Immunopharmacology*. doi: 10.1016/j.intimp.2007.08.013.

Yaltirak, T., Aslim, B., Ozturk, S., and Alii, H. (2009). Antimicrobial and antioxidant activities of *Russula delica* Fr. *Food and Chemical Toxicology*, **47**, 2052–2056.

Yang, B-K., Gu, Y-A., Jeong, Y-T., Jeong, H., and Song, C-H. (2007). Chemical characterics and immune-modulating activities of exo-biopolymers produced by *Grifola frondosa* during submerged fermentation process. *International Journal of Biological Macromoleculaes*, **41**, 227–233.

Yang, J., Zhan, W., Shi, P., Chen, J., Han, X., and Wang, Y. (2005). Animal and IN VITRO Models in Human Diseases Effects of exopolysaccharide fraction (EPSF) from a cultivated *Cordyceps sinensis* fungus on c-Myc, c-Fos, and VEGF expression in B16 melanoma-bearing mice. *Pathology – Research and Practice*, **201**, 745–750.

Yang, C-M., Zhou, Y-J., Wang, R-J., and Hu, M-L. (2009). Anti-angiogenic effects and mechanisms of polysaccharides from *Antrodia cinnamomea* with different molecular weights. *Journal Ethnopharmacology*, **123**, 407–412.

Ye, M., Luo, X., Li, L., Shi, Y., Tan, M., Weng, X., Li, W., Liu, J., and Cao, Y. (2007). Grifolin, a potential antitumor natural product from the mushroom *Albatrellus confluens*, induces cell-cycle arrest in G1 phase via the ERK1/2 pathway. *Cancer Letters*, **258**, 199–207.

Youguo, C., Zongji, S., and Xiaoping, C. (2009). Modulatory effect of *Ganoderma lucidum* polysaccharides on serum antioxidant enzymes activities in ovarian cancer rats. *Carbohydrate Polymers*, doi:10.1016/j.carbpol.2009.03.030

Zhang, L., Yang, M., Song, Y., Sun, Z., Peng, Y., Qu, K., and Zhu, H. (2009). Antihypertensive effect of 3,3,5,5-Tetramethyl-4-Piperidone: A new compound extracted from *Marasmius androceus*. *Journal of Ethnopharmacology*, **123**, 34–39.

Zheng, R., Jie, S., Hanchuan, D., and Moucheng, W. (2005). Characterization and immunomodulating Activities of Polysaccharide from *Lentinus edodes*. *International Immunopharmacology*, **5**, 811–820.

Zhu, X-L. Chen, A-F., and Lin, Z-B. (2007). *Ganoderma lucidum* polysaccharides Enhance the Function of Immunological Effector Cells In Immunosuppressed Mice. *Journal of Ethnopharmacology*, **111**, 219–226.

8

Baublis A. J., Lu C., Clydesdale F. M., and Decker E. A. (2000). Potential of wheat-based breakfast cereals as a source of dietary antioxidants. *Journal of the American College of Nutrition*, **19**, 308–311.

Benzie I. F. and Strain J. J. (1996). The Ferric Reducing Ability of Plasma (FRAP) as a measure of "Antioxidant Power": The FRAP assay. *Analytical Biochemistry*, **239**, 70–76.

Huang, M. T., Ho, C. T., and Lee, C. Y. (1992). Phenolic compounds in food and their effects on health: II. *Antioxidants and Cancer Prevention*. Washington: ACS Symposium Series 507.

Kühnau, J. (1976). The flavonoids: A class of semi-essential food components: Their role in human nutrition. *World Review of Nutrition and Dietetics*, **24**, 117–191.

Luthria, D. L. and Pastor-Corrales, M. A. (2006). Phenolic acids content of fifteen dry edible bean (*Phaseolus vulgaris* L.) varieties. *Journal of Food Composition and Analysis*, **19**, 205–211.

Mohamadi, A., Shahin, R., Emam-Djomeh, Z., and Keyhani, A. (2008). Kinetic models for colour changes in kiwifruit slices during hot air drying. *World Journal of Agricultural Sciences*, **4**(3), 376–383.

Shahidi, F. and Wanasundra, P. K. (1992). Phenolic antioxidants. *Critical Reviews in Food Science Nutrition*, **32**, 67–103.

Vansteenkiste, E., Babot, C., Rouau, X., and Micard, V. (2004). Oxidative gelation of feruloylated arabinoxylan as affected by protein. Influence on protein enzymatic hydrolysis. *Food Hydrocolloids*, **18**, 557–564.

Verschuren, P. M. (2002). Functional foods: Scientific and global perspectives. *British Journal of Nutrition*, **88**, S125–130.

Index

For Product Safety Concerns and Information please contact our EU representative GPSR@taylorandfrancis.com
Taylor & Francis Verlag GmbH, Kaufingerstraße 24, 80331 München, Germany

www.ingramcontent.com/pod-product-compliance
Ingram Content Group UK Ltd.
Pitfield, Milton Keynes, MK11 3LW, UK
UKHW021508240926
475702UK00035B/654
* 9 7 8 1 9 2 6 8 9 5 0 1 7 *